# Statistics with Stata® 3

**Lawrence C. Hamilton**
*University of New Hampshire*

**Duxbury Press**
*An Imprint of Wadsworth Publishing Company*
Belmont, California

*This book is printed on acid-free recycled paper*

**Duxbury Press**
*An Imprint of Wadsworth Publishing Company*
A division of Wadsworth, Inc.

Printed in the United States of America

1 2 3 4 5 6 7 8 9 10—97 96 95 94 93

Library of Congress Cataloging-in-Publication Data

Hamilton, Lawrence C.
    Statistics with Stata 3 / Lawrence C. Hamilton.
        p.      cm.
    Includes bibliographical references and index.
    ISBN 0-534-18918-0 (manual only). — ISBN 0-534-18920-2 (manual
with 3.5-inch IBM disk). — ISBN 0-534-18919-9 (manual with
Macintosh disk)
    1. Stata    2. Social sciences—Statistical methods—Computer
programs.   3. Social sciences—Statistics—Graphic methods—Computer
programs.   I. Title.   II. Title: Statistics with Stata three.
HA32.H363    1993
519.5'0285'5369—dc20                                         92-39339

*Stata® is available for a variety of computers, including DOS, OS/2, Macintosh, and Unix systems. A degree-granting institution can obtain a free site license for the Student Version of Stata if it already owns or purchases five copies of the Professional Version. For details, contact:*

Computing Resource Center, Santa Monica, CA
Telephone: 310/393-9893   800/782-8272 800 (STATAPC)
Fax: 310/393-7551

## Acknowledgments

I must start by acknowledging the indispensable contributions of Computing Resource Center's William Gould, who designed an elegant program and helped me understand it. Joseph Hilbe, editor of the *Stata Technical Bulletin*, also contributed heavily to this book. John Kimmel and John Moroney at Brooks/Cole provided early support, helping to get this project off the ground. Stan Loll assumed editorship under difficult conditions, and saw the work through to completion. J. Theodore Anagnoson, Richard DeLeon, and William Rogers all offered key suggestions. Stephanie Chasteen took the manuscript out for a test drive, and Patricia Branton of CRC did fine work as copy editor.

Essential help came from people who freely shared their data, notably Susan Baker (Nevada highway fatalities, Chapter 6); Sally Ward (student survey, Chapter 5); and Peggy Plass and Kirk Williams (urban homicides, Chapter 3). Warren Hamilton brought several intriguing datasets to my attention. At the University of New Hampshire, I benefited from the support of Jennifer Bakke, Petr Brym, Angele Cook, Joseph Danahy, Cynthia M. Duncan, Sean Lauer, Betty LeCompagnon, Ernst Linder, Dennis Meadows, Stuart Palmer, Deena Peschke, Murray Straus, Sally Ward, and Terri Winters. The curiosity of students at Deering, Kiana, Kivalina, and other Northwest Arctic schools supplied last-minute encouragement.

I dedicate this book to Leslie, who made it all possible and occasionally even fun.

Statistics Editor: **Stan Loll**
Editorial Assistant: **SoEun Park**
Production Editor: **Carol Lombardi**
Managing Designer: **Andrew H. Ogus**
Cover Design: **Stuart Paterson**
Print Buyer: **Diana Spence**
Printer: **Malloy Lithographing, Inc.**

# Contents

# 1
# Introduction to Stata

*Statistics with Stata* shows how to use Stata version 3 as a power tool for data analysis. Some copies of the book also include disks containing Student Stata and numerous example datasets. I use these examples to illustrate basic analytical procedures, adding brief interpretive notes as needed. *Statistics with Stata* is not itself a textbook, but could provide a practical supplement to any applied statistics text.[1]

The first seven chapters of this book cover topics generally corresponding to a first course in statistics: from data entry through elementary univariate and bivariate techniques, ending with multiple regression. The remaining chapters contain how-to material well beyond the scope of introductory courses, but provide essential tools for original research. Of course, a 200-page book cannot hope to cover all of Stata's features—only some of the most important. Stata's online **help** facility provides further details, including command syntax and full listing of options. For complete documentation, consult the three-volume *Stata Reference Manual* that comes with Professional Stata.

Stata itself is a full-featured statistical program for DOS (IBM-compatible), Macintosh, OS/2, or Unix computers. Compared with other statistical programs, Stata is relatively lean, quick, and easy to learn. It has a consistent, flexible command syntax, plus optional menus and extensive online help for assistance. Advanced users appreciate the built-in programming language, which has given rise to a journal (*Stata Technical Bulletin*) and update disks that bring new capabilities every two months.[2] Leading-edge statistical developments such as quantile regression thus quickly become part of Stata, as do simpler innovations for teaching or special purposes.

No program can do everything, but the things Stata does, it does well. Its graphical and regression capabilities deserve particular mention, and fit nicely together. If Stata's capabilities encompass the main techniques of your field, you may find that the experience of using Stata compares with using older mainframe-derived programs rather like driving a sports car compares with driving a bus.

Under DOS, Stata version 3 requires a floppy disk drive (3.5" or 5.25"), a hard disk drive, and at least 640K of memory. It automatically configures itself to use the available graphics display and expanded memory, and runs happily under multitasking programs like Windows or Desqview. A 32-bit version, Intercooled Stata, requires more hardware (386SX or better CPU, math coprocessor, 4M RAM) but runs faster and can analyze huge datasets. On a Macintosh, regular Stata requires 2M and Intercooled Stata 4M of memory. All OS/2 and Unix versions of Stata are comparable to Intercooled Stata.

---

[1] Two texts that fit well with *Statistics with Stata* are:
*Modern Data Analysis: A First Course in Applied Statistics* (Hamilton 1990)
*Regression with Graphics: A Second Course in Applied Statistics* (Hamilton 1992a).

[2] To obtain reprints, subscriptions, or submission guidelines for *Stata Technical Bulletin*, contact Computing Resource Center (CRC), 1640 Fifth Street, Santa Monica, CA 90401 (telephone 800 STATAPC).

This book may include disks with the DOS or Macintosh Student Stata program, plus instructions on installation. (Student Stata versions do not yet exist for OS/2 or Unix.) Unlike student versions of other microcomputer programs, Student Stata is not just courseware or a small subset of its professional cousin. The main differences between Student Stata and Professional Stata are:

1.    Datasets in Student Stata are limited to 160 observations by 25 variables. With enough RAM, Professional Stata handles datasets up to 32,000 by 254 (for Unix and Intercooled versions, limited only by RAM, 500,000 by 2,000 or more is possible).

2.    Student Stata can display graphs onscreen, but not save them to disk or print them. Professional Stata can save, combine, or print high-resolution graphs, and translate them into formats that are readable by other programs. A companion program, Stage (**Stata Graphics Editor**), edits and enhances Professional Stata graphs.

3.    Student Stata will not analyze cross-product datasets. Professional Stata has this capability, which is unlikely to be missed by beginning users.

4.    Student Stata includes example datasets mentioned in this book; Professional Stata does not. (The datasets contained on Student Stata disks can be used with Professional Stata, however; see the separately printed installation instructions.)

In most other respects, Student and Professional Stata are the same. Consequently, whatever you learn and any data or programs you create using Student Stata should work equally well with Professional Stata—simplifying the transition from student to researcher.

## A Typographical Note

In this book I employ certain typographical conventions:

> **commands typed by the user** appear in a bold, typewriter-style font. End all commands by hitting the Enter key (on some keyboards, marked Return). If a command goes beyond one line, keep typing—the display will wrap, and the command does not execute until you hit Enter.

> Stata output, programs, and program names also appear in a typewriter font.

> *variable names*, *file names*, and *hard disk directory names* are in italics. Such names might be incorporated within a Stata command, but writing them in italics emphasizes that they are arbitrary—other appropriate names could be substituted. For example, to find the correlation between two variables named *elev* (elevation) and *wind* (wind power):

```
. correlate elev wind
(obs=21)
        |     elev      wind
--------+------------------
   elev|   1.0000
   wind|   0.2563    1.0000
```

Finally, angle brackets denote special keys like <Enter>, <F10>, or <PgUp>.

## Starting and Exiting Stata (DOS version)

To begin a Stata session from DOS, either:

1.    if you earlier added c:\stata (or c:\stustata for Student Stata) to the path statement in your autoexec.bat file, just type **stata** (or **stustata**);

2.   otherwise, make Stata's directory current, then type **stata** (or **stustata** for Student
     Stata).  For example, if you followed the default installation, placing Student Stata in
     a directory called *stustata* on drive C, you could type:

```
C:\>cd \stustata
C:\STUSTATA>stustata
```

Once Stata takes over, you will see a dot prompt, indicating Stata awaits your command.

   To end a Stata session and return to DOS, simply type:

```
. exit
```

Sometimes Stata won't let you go:

```
. exit
no; data in memory would be lost
r(4);
```

This error message warns that the data currently in memory have not been saved to a disk file.
After seeing this warning, you have two choices:

1.   Exit without saving the data, by typing:

```
. exit, clear
```

2.   Save the data and then exit, for example by typing

```
. save yourdata
. exit
```

   or

```
. save yourdata, replace
. exit
```

With the  **replace** option, Stata overwrites whatever used to be in file *yourdata*.  The
**replace** option updates a previously saved dataset.  To save a new dataset, simply type **save**
followed by a new filename.

   Commands like **save**, that refer to files on disk, understand DOS designations for disk
drives and directories.  If you wish to save *yourdata* on the A drive, for instance:

```
. save a:yourdata
```

or in the *stustata* directory on the C drive:

```
. save c:\stustata\yourdata
```

Unless told otherwise, Stata assumes file commands refer to the current directory.  You can
change this default to something else with the **set prefix** command.  For example, if we type

```
. set prefix c:\data
```

then subsequent commands like

```
. save yourdata
```

will be understood as if we had really typed

```
. save c:\data\yourdata
```

Some DOS commands like `del` (delete), `copy`, or `dir` (list files in directory) can be used without leaving Stata. Precede these commands with an exclamation mark. For example, to back up all datasets (`.dta` files) from the hard disk directory *c:\stustata* onto a floppy disk in drive A, type:

```
. !copy c:\stustata\*.dta a:
```

## Starting and Exiting Stata (Macintosh version)

To begin a Stata session on the Macintosh, open the *StuStata* folder and double-click on `Stata.do`. The Macintosh will automatically find Stata. Once Stata takes over, you will see the " . " prompt, indicating that Stata awaits your command. To end a Stata session and return to the Finder, simply click on the close box.

Sometimes, you will get a dialog box informing you that Stata's data area has been changed, but not saved. You then have three options:

1.  If you want to exit without saving your the data, click on the box `Exit Anyway`.
2.  If you do want to save the data, click on the box marked `Save`. Stata then asks for a file name, under which to save your data.
3.  If you decide you do not wish to exit after all, return to Stata by clicking on the `Don't Exit` box.

Throughout this book, I use DOS filenaming conventions. When I say to type **use** *c:\stustata\wind*, for instance, you can type **use wind**. The dataset (*wind.dta*) is in the *StuStata* folder. Macintosh users can also enter Stata by finding the dataset, say *wind.dta*, in the *StuStata* folder and double-clicking on it. If you do this, though, you will not see Student Stata's opening messages telling how to use it.

## A Sample Stata Session

The following brief session illustrates how Stata works. We begin as usual by making Stata's directory current, then typing **stata** (or **stustata**). Next we retrieve some data from a study evaluating the potential for wind-generated electricity in Jamaica (Chen et al. 1990). These data exist in a Stata-format disk file called *wind.dta*, provided with Student Stata:

```
. use c:\stustata\wind
(Jamaica Wind (Chen et al. 1990))

. describe
Contains data from c:\stustata\wind.dta
  Obs:      22 (max=   2620)              Jamaica Wind (Chen et al. 1990)
  Vars:      4 (max=     99)
Width:      20 (max=    200)
   1. site            str15   %15s              Station
   2. side            byte    %8.0g    slbl     Which Side of Island?
   3. elev            int     %8.0g             Elevation in Meters
   4. wind            int     %8.0g             Avg. Wind Power/meter^2
Sorted by:
```

**describe** briefly characterizes the data. We have 22 observations and 4 variables. Stata also reports the maximum number of observations and variables it could accommodate, as currently configured (see Dataset Size in Chapter 2). Width refers to the amount of memory (in bytes) currently available for each observation. Type **help maxvar** for more details about maximum observations, variables, and width (most important with large datasets). Other elements of **describe**'s output include:

| data label: | "Jamaica Wind (Chen et al. 1990)" |
| variable names: | *site, side, elev,* and *wind* |
| variable types: | *site* is a string variable, up to 15 characters long (str15). *side* is a byte-format variable; *elev* and *wind* are both integers (see page 17 or type **help datatypes** for more about Stata variable formats). |
| value labels: | the values of *side* are labeled, with labels stored as *slbl.* |
| variable labels: | Variable labels such as "Elevation in Meters" or "Avg. Wind Power/meter^2" explain the meaning of variable names. (Incidentally, ^ is a standard computer symbol for raise to power, so this variable label means "wind power per square meter.") |

To actually see the data in memory, we may want to **list** them:

. **list**

| | site | side | elev | wind |
|---|---|---|---|---|
| 1. | Bodles | south | 30 | 31 |
| 2. | Crawford | south | 30 | 34 |
| 3. | Discovery Bay | north | 0 | 200 |
| 4. | Fairy Hill | north | 76 | 134 |
| 5. | Flagaman | south | 60 | 200 |
| 6. | Folly Point | north | 0 | 93 |
| 7. | Fullerswood | south | 0 | 14 |
| 8. | Galina | north | 0 | 236 |
| 9. | Hellshire | south | 0 | 198 |
| 10. | Hillside | south | 30 | 140 |
| 11. | Manchoneal | north | 6 | 50 |
| 12. | Manley | south | 0 | 188 |
| 13. | Mason River | north | . | 24 |
| 14. | Morant Point | south | 0 | 193 |
| 15. | Munro | south | 792 | 237 |
| 16. | Passley Gardens | north | 15 | 70 |
| 17. | Pimento Hill | north | 305 | 200 |
| 18. | Rowlandsfield | south | 305 | 55 |
| 19. | Sangster | north | 0 | 145 |
| 20. | Spur Tree | south | 610 | 199 |
| 21. | Vinery | north | 457 | 70 |
| 22. | Yallahs | south | 60 | 100 |

Note the period (.) for Mason River's elevation; this is Stata's "missing value" symbol.

**list**, by itself, asks Stata to list all the data in memory. Typing

. **list** *site wind*

would have produced a listing only for the variables *site* and *wind*.

Data analysis might begin with summary statistics:

. **summarize**

| Variable | Obs | Mean | Std. Dev. | Min | Max |
|---|---|---|---|---|---|
| site | 0 | | | | |
| side | 22 | .4545455 | .5096472 | 0 | 1 |
| elev | 21 | 132.1905 | 228.3899 | 0 | 792 |
| wind | 22 | 127.7727 | 75.44719 | 14 | 237 |

Because *site* is a string variable, it has no mean or standard deviation and **summarize** treats all its values as missing. *side*, on the other hand, is a labeled numeric variable (0="south", 1="north"). **summarize** notes *side*'s minimum (0), maximum (1), and calculates a mean and standard deviation, though when we **list**ed the data we saw *side*'s value labels "south" and "north."

Does wind power correlate with elevation? To find out, type:

```
. correlate elev wind
(obs=21)
        |     elev     wind
--------+------------------
   elev|   1.0000
   wind|   0.2563   1.0000
```

We see only weak correlation. But perhaps the wind/elevation correlation depends on the side of the island? The **by** prefix repeats the analysis for each value of *side*:

```
. sort side
. by side: correlate elev wind

-> side=    south  (obs=12)
        |     elev     wind
--------+------------------
   elev|   1.0000
   wind|   0.4015   1.0000

-> side=    north  (obs=9)
        |     elev     wind
--------+------------------
   elev|   1.0000
   wind|  -0.1086   1.0000
```

Southern sites exhibit a moderate positive correlation ($r = .4015$), while the north side shows virtually none ($r = -.1086$). Other variables such as prevailing wind directions, tree cover, and local topography must account for most of the variation in wind power.

This simple example illustrates interactive data analysis, which Stata makes easy: we look at our data, try some analyses, then explore further possibilities suggested by the results. Learning about the data takes place through a process of exploration and discovery.

To leave Stata, type **exit, clear**. If you ran Stata from the *stustata* directory of your C drive, the sample session just seen required this sequence of commands:

```
C:\>cd \stustata
C:\STUSTATA>stustata
. use c:\stustata\wind
. describe
. list
. summarize
. correlate wind elev
. sort side
. by side:  correlate wind elev
. exit, clear
C:\STUSTATA>
```

If we used Professional instead of Student Stata, the session could look slightly different:

```
C:\>stata
. addpath + c:\stustata
. use wind
. describe
. list
. summarize
. correlate wind elev
. sort side
. by side:  correlate wind elev
. exit, clear
C:\>
```

## Printing and Saving Results

Suppose we wanted a printout or "hard copy" of our sample Stata session. If the computer is connected to a printer, `log using prn:` usually works:

```
. log using prn:
. use c:\stustata\wind
. describe
. list
. summarize
. correlate wind elev
. log close
. exit, clear
```

All our commands and Stata's responses, from `log using prn:` to `log close`, should appear on the printout. (Graphs will not appear in a log file printout. See Chapter 3 for instructions on printing graphs.)

A Stata log, once begun, can be turned off and on any number of times, allowing selective printing. For example, to print only the data description and summary statistics from the previous example:

```
. log using prn:
. log off
. use c:\stustata\wind
. log on
. describe
. log off
. list
. log on
. summarize
. log off
. correlate wind elev
. log close
. exit, clear
```

Note the distinction between `log off` (temporary) and `log close` (permanent).

`log using prn:` copies all onscreen text to the printer. Alternatively, we might store results in a disk file. The command

```
. log using filename
```

sends all subsequent text output to *filename*`.log` in the current directory. (Macintosh users, in fact, must use this approach; they can later print the log file by double-clicking on it.) Some variations on this command:

```
. log using a:monday1
```

(output goes to file *monday1*`.log` , on disk in drive A).

```
. log using c:\stustata\example
```

(output goes to file *example*`.log` , in the *stustata* directory on drive C). `log off`, `log on`, and `log close` work the same way regardless of whether the destination is a printer or disk file.

There are several things we can do with a log file on disk. Such files are in ASCII (text; pronounced "ask-ee") format, and so can be edited or incorporated into a larger document by almost any word processor. We can always print a disk log file, by telling DOS to copy it to the printer:

```
C:\>copy a:monday1.log prn:
```

Repeat this command to print additional copies. The original disk version is unaffected.

Besides recording results, log files record any error messages or mistakes. This sounds embarrassing, but it is also valuable. From time to time you may get stuck and need help figuring out what you are doing wrong. When consulting the local Stata guru, always try to bring along your disks and a log transcript of the problem. If you can't get a transcript, at least copy down the exact error message. It is much easier to help someone who can provide detailed information, rather than a generic complaint like "I tried it and it didn't work."

Beginners often become confused about the difference between Stata log files and data files. The command

```
. log using filename
```

begins a log file called *filename*.log , that will contain a literal transcript of all text input and output appearing on your screen from now until you interrupt it with **log off** or terminate it with **log close**. This transcript can be printed or edited by a word processor, but it is basically inert—it is merely a record of what happened. Stata cannot read or analyze a log file. The command

```
. save filename
```

or

```
. save filename, replace
```

saves a Stata-format data file called *filename*.dta . This data file contains information that Stata can read, analyze, edit, or add to any number of times. It does not contain the analyses themselves. In the Stata sample session shown earlier, file *wind.dta* is a Stata-format data file.

## Menus (DOS only)

The sample Stata session was conducted entirely with short commands such as **describe**, **list**, and **correlate**. Most Stata commands can be abbreviated to their first few letters, making them even shorter. To new users, however, the greatest challenge is not typing commands, but remembering them.

Alternatively, we could control the session through menus rather than command lines. To invoke menus at any point in a Stata session, type:

```
. menu
```

or hit the <F10> key. (Macintosh, OS/2, Unix, and Intercooled versions of Stata do not have a menu command.) The screen then fills with a list of possible choices, with instructions along the bottom. Use arrow or <Tab> keys to move the cursor to the command you want, then press <F10>. This brings up a form with blanks to complete. Some blanks refer to advanced options, or restrict the analysis to certain cases—not all of these need be filled out. When you have filled in the minimum necessary information, press <F10> and the command executes.

Experienced users, like diners in a familiar restaurant, may prefer to bypass the menus and just ask for what they want. You can switch back and forth between menu and nonmenu modes, invoking menus by typing **menu** and suppressing them by hitting the escape <Esc> key. Throughout this book I employ command-line mode, which is more compact and easier to demonstrate. Many of the same commands could be issued via menus, however.

# Using the Keyboard

Most of the keys on your keyboard function the same way in Stata as they do in other programs. There are a few important exceptions, keys that have special meanings in Stata. Notation such as <Esc> means tap the escape key, rather than literally typing the three letters E-s-c. <Ctrl><Break> means hold down the control key <Ctrl> and tap <Break>.

<Ctrl><Break> (DOS, OS/2)    <Ctrl><C> (Unix)    <Apple><.> (Macintosh) stops the execution of a Stata command. This helps when you realize your last command was erroneous, or will produce too much output or take too much time.

<Esc> (DOS, OS/2)    <Ctrl><U> (Unix and Macintosh) deletes the current command line, so you can start over.

<PgUp> (DOS, OS/2 and some Unix)    <Ctrl><R> (other Unix and Macintosh) recalls the previous command. This can be tremendously useful. Hitting <PgUp> twice recalls the command before that, and so on for many previous commands. Once recalled, a command can be resubmitted as is, by hitting <Enter>, or edited with the arrow, backspace, insert <Ins>, or delete <Del> keys. This saves typing and effort in correcting mistakes, repeating similar analyses with small changes, or building graphs.

<PgDn> (DOS, OS/2, and some Unix)    <Ctrl><B> (other Unix and Macintosh) recalls the command after the one currently showing (if that command was brought back by <PgUp>).

<Tab> (DOS, OS/2) moves from one field to the next in menu mode.

Some function or F-keys also have special meanings in Stata:

```
<F1>    help
<F2>    #review
<F3>    describe
<F7>    save
<F8>    use
<F10>   menu
```

For instance, tapping the <F3> key means the same thing as typing **describe**.

# Online Help

Stata offers extensive online help facilities. At any time during a Stata session, type

`. help`

(or tap <F1>) to see a list of topics for which help is available. If you already know you need help with the **summarize** command, for instance, you could instead type

`. help summarize`

Stata responds with the complete syntax for this command, including all options, together with a brief explanation and examples. Similarly, for details about the **correlate** command, type:

`. help correlate`

and so forth.

Stata's **help** feature is essentially an online manual, providing more complete information about commands and options than this book. You can learn much about Stata by browsing through **help**.

Each chapter of this book ends with an Also Type **help** section, listing relevant topics for which online help is available. If neither *Statistics with Stata* nor **help** answers your questions, you need the full documentation of the *Stata Reference Manual*.

## Also Type **help**

| | |
|---|---|
| **by** | repeat operation for categories of a variable |
| **correlate** | Pearson correlation |
| **datatypes** | variable types and storage formats |
| **describe** | list variables and other information about current dataset |
| **exit** | leave Stata (to exit without saving data: **exit, clear**) |
| **format** | variable display format (decimal place, rounding, etc.) |
| **help** | online help available |
| **key** | special key functions |
| **label** | text labels for variables, values, or dataset |
| **list** | list the data in memory |
| **log** | save or print Stata *output* (except graphs) |
| **save** | save Stata *dataset* (to write over earlier version: **save, replace**) |
| **set** | set program parameters |
| **shell** | temporarily leave Stata, go to DOS |
| **sort** | re-order cases according to values of listed variable(s) |
| **summarize** | find means, standard deviations, and other statistics |
| **use** | retrieve previously-**save**d Stata dataset |

# 2
# Data

This chapter shows how to create a Stata-format dataset. We can start by typing in the data directly or by reading in data already saved by another program. Once the data are in Stata, we can label, edit, add variables or cases, and combine or simplify datasets as desired. Saving the dataset in Stata format makes it simple to retrieve this information later for further analysis.

Stata works like a word processor: it keeps the data in memory. When we enter data, it goes into memory; no permanent (disk) copy exists until we **save** it. Similarly, when we **use** a previously-**save**d dataset, we load it into memory. Any changes we make to the data affect only the copy in memory; they do not become permanent unless we type **save, replace** .

## Typing in a New Dataset

Table 1 lists data on the nine major planets of our solar system, from Beatty, O'Leary, and Chaikin (1981):

Table 1: Physical Characteristics of Nine Major Planets

| Planet | Distance from Sun (million km) | Equatorial Radius (km) | Ringed? | No. of Moons | Mass (kg) |
|--------|-------------------------------|------------------------|---------|--------------|-----------|
| Mercury | 57.9 | 2439 | none | 0 | 3.30e+23 |
| Venus | 108.2 | 6050 | none | 0 | 4.87e+24 |
| Earth | 149.6 | 6378 | none | 1 | 5.98e+24 |
| Mars | 227.9 | 3398 | none | 2 | 6.42e+23 |
| Jupiter | 778.3 | 71900 | rings | 16 | 1.90e+27 |
| Saturn | 1427.0 | 60000 | rings | 17 | 5.69e+26 |
| Uranus | 2869.6 | 26145 | rings | 15 | 8.66e+25 |
| Neptune | 4496.6 | 24750 | rings | 8 | 1.03e+26 |
| Pluto | 5900.0 | 1550 | none | 1 | 1.10e+22 |

To type raw data directly into Stata, we begin with the **input** command, followed by a list of names (up to eight characters each) for the variables we wish to enter. For example, to enter distance from sun, radius, and rings (0=no rings, 1=has rings) for the first six planets:

```
. clear
. input dsun radius rings

         dsun        radius        rings
  1. 57.9 2439 0
  2. 108.2 6050 0
  3. 149.6 6378 0
  4. 227.9 3398 0
  5. 778.3 99900 1
  6. 1427 60000 1
  7. end
```

After **input**, Stata prompts with case numbers "1.", "2.", and so on. Input columns need not line up, but there must be at least one space between each variable value. We type **end** to stop data entry and return to Stata's usual "." prompt.

Stata represents missing values with a period.  For example, suppose we did not know Saturn's radius.  Then the 6th line of our input would look like this:

```
6.  1427 . 1
```

Do not leave missing values blank.  Always type them in as a period, preceded and followed by at least one space.

After typing in data, we should name the dataset and save it to disk.  The command

```
. save c:\stustata\planets
file c:\stustata\planets.dta saved
```

creates a Stata-format data file called *planets.dta*.  Stata automatically adds a `.dta` extension to the filename, to mark this as a data file.

A **save**d dataset can be used in subsequent Stata sessions.  We can later add further information to this file and save the updated version by typing

```
. save c:\stustata\planets, replace
```

or simply

```
. save, replace
```

Reminder:  Stata will not let us **use** (retrieve) a new dataset while another, un**save**d, dataset is already in memory.  If we have been working with *planets.dta* and now want to retrieve *wind.dta*, we must either first **save** *planets.dta*:

```
. save c:\stustata\planets, replace
. use c:\stustata\wind
```

or, if we don't want to save *planets.dta* (or whatever is in memory), just type

```
. use c:\stustata\wind, clear
```

To clear data from memory without reading in a new dataset, type

```
. clear
```

## Selecting Cases and Correcting Mistakes

You may have noticed that I mistakenly made Jupiter too large, typing its radius as 99900 kilometers instead of the correct 71900.  Such mistakes are easily fixed:

```
. replace radius = 71900 in 5
(1 real change made)
```

**replace** changes the value of a variable.  **in 5** qualifies this operation so it applies only to the 5th case.  (Had we typed simply **replace radius = 71900** , without the **in 5** qualifier, that would change <u>all</u> *radius* values to 71900.)  After correcting any mistakes, or otherwise improving the data, save the newer version to disk by typing:

```
. save c:\stustata\planets, replace
```

or simply

```
. save, replace
```

(Stata will assume that the name has not changed.)

**in** qualifiers can be added to any Stata editing or analysis command, restricting that command to a subset of the cases.  For example, to **list** only the 3rd case in a dataset:

. list in 3

To **list** the 19th through 24th cases:

. list in 19/24

To **summarize** the last 5 cases (especially useful after **sort**):

. summarize in -5/1

The notation **-5** refers to the 5th-from last case, and a lower-case letter "l" denotes the last case. The numeral **1**, which looks so similar to letter **l**, means something quite different to Stata: the first case. Stata likewise distinguishes between the letter **o** and the numeral **0**, so we need to read and type carefully to avoid confusing these symbols.

Another way to correct mistakes employs an **if** qualifier:

. replace *radius* = 71900 if *radius*==99900

Whereas **in** restricts an operation to certain case numbers, **if** restricts the operation to any cases meeting the specified criteria. Note the two different kinds of equality in this command:

> **replace *radius* = 71900**    calls for an algebraic operation: "make the left-hand side of this equality the same as what is on the right"

On the other hand,

> **if *radius*==71900**    calls for a comparison: "check to see whether it is true that the left-hand side is the same as the right"

The double equals sign **==** denotes a comparison operator. Stata's comparison operators are:

- **==** equal to
- **>** greater than
- **<** less than
- **>=** greater than or equal to
- **<=** less than or equal to
- **~=** not equal to (!= also works)

Like **in**, the qualifier **if** can restrict any data-editing or analytical command. For example:

. list if *radius* < 6400

or

. summarize *radius* if *dsun* >= 1000

Two or more comparison operations can be combined in a single **if** expression by using logical operators. Stata's logical operators are:

- **&** and
- **|** or
- **~** not (! also works)

For example, to find the mean radius of planets that are between 100 and 500 million kilometers from the sun:

. summarize *radius* if *dsun* > 100 & *dsun* < 500

| Variable | Obs | Mean | Std. Dev. | Min | Max |
|---|---|---|---|---|---|
| radius | 3 | 5275.333 | 1634.069 | 3398 | 6378 |

We find three such planets; their mean radius equals 5275.333 kilometers.

## Adding Variables or Cases

So far, we have entered, corrected, and saved data on three variables and six cases (planets). Suppose these initial *planets.dta* data are in memory, and now we want to add more cases—such as the three outer planets, Uranus, Neptune, and Pluto. To begin adding cases, just type `input`; Stata responds by prompting for the next case number:

```
. input

          dsun      radius      rings
7.  2869.6 26145 1
8.  4496.6 24750 1
9.  5900 1550 0
10. end

. save, replace
```

We now have data on the nine major planets.

Suppose we wish to add more variables. Again use `input`, this time specifying the new variables' names. To add variables indicating the number of known moons and mass:

```
. input moons mass

      moons      mass
1.  0   3.3e+23
2.  0   4.87e+24
3.  1   5.98e+24
4.  2   6.42e+23
5.  16  1.9e+27
6.  17  5.69e+26
7.  15  8.66e+25
8.  8   1.03e+26
9.  1   1.1e+22

. save, replace
```

The *mass* values are very large numbers, here expressed in the computer version of scientific notation. Mercury's mass appears as 3.3e+23, meaning

$$3.3 \times 10^{23} = 330,000,000,000,000,000,000,000$$

kilograms. Very small numbers (close to zero) are written with negative exponents. For example, 1.2e–09 means

$$1.2 \times 10^{-9} = .0000000012$$

Stata automatically switches to scientific notation where necessary in its output, and it can read input in either scientific or ordinary notation. Consult the *Stata Reference Manual* for details about how Stata stores numbers internally or `help format` regarding how it displays them.

## String Variables

All the variables we typed in so far have been numerical. Stata can also handle string variables, which include any sequence of up to 80 characters and spaces. For example, we might want to type in the planets' names. The `input` command must notify Stata that the next-named variable will be a string variable, and specify its maximum length. The planets have names of 7 letters or less, so we tell Stata to expect a `str7` variable named *planet*:

```
. input str7 planet

      planet
1.  "Mercury"
```

```
2.  "Venus"
3.  "Earth"
4.  "Mars"
5.  "Jupiter"
6.  "Saturn"
7.  "Uranus"
8.  Neptune
9.  Pluto
```

A 27-character string variable would be **str27**, and so on up to **str80**. Quotations (double or single) are necessary if the strings include any blanks or special characters; otherwise (as illustrated with Neptune and Pluto above), quotes are optional.

## Labeling Data, Variables, and Values

We have created a dataset that looks like this:

```
. describe

Contains data from c:\stustata\planets.dta
  Obs:      9 (max=   2620)
  Vars:     6 (max=     99)
Width:     21 (max=    200)
    1. dsun         float    %9.0g
    2. radius       float    %9.0g
    3. rings        float    %9.0g
    4. moons        float    %9.0g
    5. mass         float    %9.0g
    6. planet       str7     %9s
Sorted by:
Note:  Data has changed since last save
```

```
. list

          dsun     radius      rings      moons        mass     planet
1.        57.9       2439          0          0    3.30e+23    Mercury
2.       108.2       6050          0          0    4.87e+24      Venus
3.       149.6       6378          0          1    5.98e+24      Earth
4.       227.9       3398          0          2    6.42e+23       Mars
5.       778.3      71900          1         16    1.90e+27    Jupiter
6.        1427      60000          1         17    5.69e+26     Saturn
7.      2869.6      26145          1         15    8.66e+25     Uranus
8.      4496.6      24750          1          8    1.03e+26    Neptune
9.        5900       1550          0          1    1.10e+22      Pluto
```

Although the basic information is here, this "bare bones" dataset contains no internal documentation. Adding labels will make the variables' definitions much easier to remember. Stata datasets may contain three kinds of labels:

| | |
|---|---|
| **label data** | describes the dataset as a whole |
| **label variable** | describes a variable |
| **label values** | describes individual values |

For example, to label the *planets* dataset, type

```
. label data "Solar system data"
```

To label the individual variables, type

```
. label variable dsun "Mean dist. sun, km*10^6"
. label variable radius "Equatorial radius in km"
. label variable mass "Mass in kilograms"
. label variable rings "Has rings?"
```

```
. label variable moons "Number of known moons"
. label variable mass "Mass in kilograms"
. label variable planet "Planet"
```

Although *rings* is a numerical variable, its values (0 or 1) stand for categories (no rings or has rings). We can label variable values with two commands. The first defines the labels:

```
. label define ringlbl 0 "none" 1 "rings"
```

Then we attach these labels (stored as *ringlbl*) to the values of variable *rings*:

```
. label values rings ringlbl
```

Stata will now display value labels instead of *rings'* actual numerical values whenever reasonable:

```
. tabulate rings

  Has rings?|      Freq.       Percent        Cum.
------------+-----------------------------------
      none |          5         55.56       55.56
     rings |          4         44.44      100.00
------------+-----------------------------------
     Total |          9        100.00
```

The original numerical values are still there, underneath the new labels. To see those numerical values, specify the **nolabel** option with **tabulate**:

```
. tabulate rings, nolabel

  Has rings?|      Freq.       Percent        Cum.
------------+-----------------------------------
         0 |          5         55.56       55.56
         1 |          4         44.44      100.00
------------+-----------------------------------
     Total |          9        100.00
```

The labeling effort made our dataset self-documenting. **describe** and **list** show:

```
. describe

Contains data from c:\stustata\planets.dta
  Obs:       9 (max=  2620)                    Solar system data
  Vars:      6 (max=    99)
Width:      21 (max=   200)
    1. dsun        float   %9.0g              Mean dist. sun, km*10^6
    2. radius      float   %9.0g              Equatorial radius in km
    3. rings       float   %9.0g    ringlbl   Has rings?
    4. moons       float   %9.0g              Number of known moons
    5. masskg      float   %9.0g              Mass in kilograms
    6. planet      str8    %9s                Planet
Sorted by:
Note:  Data has changed since last save
```

```
. list

        dsun     radius     rings     moons     masskg    planet
 1.     57.9       2439      none         0   3.30e+23   Mercury
 2.    108.2       6050      none         0   4.87e+24     Venus
 3.    149.6       6378      none         1   5.98e+24     Earth
 4.    227.9       3398      none         2   6.42e+23      Mars
 5.    778.3      71900     rings        16   1.90e+27   Jupiter
 6.     1427      60000     rings        17   5.69e+26    Saturn
 7.   2869.6      26145     rings        15   8.66e+25    Uranus
 8.   4496.6      24750     rings         8   1.03e+26   Neptune
 9.     5900       1550      none         1   1.10e+22     Pluto
```

The **describe** output cautions that data have changed since the last time we saved them; we just added labels. To make these changes a permanent part of the file, save the data presently in memory, replacing the previous version:

```
. save, replace
```

An aside regarding larger datasets: A single set of value labels can be attached to values of any number of variables. For example, a survey might include 20 opinion questions (I'll call them *Q1* through *Q20*), all answered with the same three-point scale: 1=disagree, 2=neutral, 3=agree. To apply labels to the values of variables *Q1-Q20*:

```
. label define agreelbl 1 "disagree" 2 "neutral" 3 "agree"
. label values Q1 agreelbl
. label values Q2 agreelbl
. label values Q3 agreelbl
```

and so on. **label define** creates the labels themselves (here called *agreelbl*), and **label values** attaches these labels to values of specific variables.

## Rearranging a Dataset: compress, order, and sort

Now, our *planets* dataset looks like this:

```
. describe

Contains data from c:\stustata\planets.dta
  Obs:       9 (max=  2620)          Solar system data
  Vars:      6 (max=    99)
Width:      21 (max=   200)
    1. dsun        float    %9.0g            Mean dist. sun, km*10^6
    2. radius      float    %9.0g            Equatorial radius in km
    3. rings       float    %9.0g    ringlbl Has rings?
    4. moons       float    %9.0g            Number of known moons
    5. mass        float    %9.0g            Mass in kilograms
    6. planet      str7     %9s              Planet
Sorted by:
```

Notations like float %9.0g and str7 %9s describe each variable's type and display format. Stata employs five different types of numeric variables:

| type | minimum | maximum | stored as |
|------|---------|---------|-----------|
| byte | −127 | 126 | 1-byte integers |
| int | −32,768 | 32,766 | 2-byte integers |
| long | −2,147,483,648 | 2,147,583,646 | 4-byte integers |
| float | $\pm 10^{-36}$ | $\pm 10^{37}$ | 4-byte reals |
| double | $\pm 10^{-99}$ | $\pm 10^{99}$ | 8-byte reals |

Floating-point (float) variables take up more memory than large integers (int), and large integers more than small integers (byte). The default type is float, which Stata automatically assigned (because we did not tell it otherwise) to the numerical variables in *planets.dta*. **compress** changes variables to more compact storage formats, where possible:

```
. compress
rings was float now byte
moons was float now byte
```

Since *rings* values are just 0 or 1, and *moons* 0–17, these two could be stored as byte variables rather than float, saving 3 bytes per value. **compress** automatically makes such changes, shrinking the dataset's size and freeing up memory or disk space. The extra memory made available by **compress** matters little with tiny datasets like *planets.dta*, but may become crucial in analyzing larger datasets.

When **compress** changes variable types, it may also change variable order in the dataset:

```
. describe

Contains data from c:\stustata\planets.dta
  Obs:        9 (max=  2620)                  Solar system data
  Vars:       6 (max=    99)
  Width:     21 (max=   200)
    1. dsun      float    %9.0g              Mean dist. sun, km*10^6
    2. radius    float    %9.0g              Equatorial radius in km
    3. mass      float    %9.0g              Mass in kilograms
    4. planet    str7     %9s                Planet
    5. rings     byte     %8.0g    ringlbl   Has rings?
    6. moons     byte     %8.0g              Number of known moons
Sorted by:
Note:  Data has changed since last save
```

We can re-order the variables ourselves, using an **order** command. For example, we might want planets' names to be the first variable in the dataset:

```
. order planet dsun radius rings moons mass

Contains data from c:\stustata\planets.dta
  Obs:        9 (max=  2620)                  Solar system data
  Vars:       6 (max=    99)
  Width:     21 (max=   200)
    1. planet    str7     %9s                Planet
    2. dsun      float    %9.0g              Mean dist. sun, km*10^6
    3. radius    float    %9.0g              Equatorial radius in km
    4. rings     byte     %8.0g    ringlbl   Has rings?
    5. moons     byte     %8.0g              Number of known moons
    6. mass      float    %9.0g              Mass in kilograms
Sorted by:
Note:  Data has changed since last save

. list

        planet     dsun    radius    rings    moons       mass
 1.     Mercury     57.9     2439     none        0    3.30e+23
 2.      Venus     108.2     6050     none        0    4.87e+24
 3.      Earth     149.6     6378     none        1    5.98e+24
 4.       Mars     227.9     3398     none        2    6.42e+23
 5.    Jupiter     778.3    71900    rings       16    1.90e+27
 6.     Saturn      1427    60000    rings       17    5.69e+26
 7.     Uranus    2869.6    26145    rings       15    8.66e+25
 8.    Neptune    4496.6    24750    rings        8    1.03e+26
 9.      Pluto      5900     1550     none        1    1.10e+22

. save, replace
```

**order** affects only the order of variables, not the order of observations. To change the order of observations, use **sort**. For example, we could **sort** the nine planets from smallest to largest radius:

```
. sort radius
. list

        planet     dsun    radius    rings    moons       mass
 1.      Pluto      5900     1550     none        1    1.10e+22
 2.     Mercury     57.9     2439     none        0    3.30e+23
```

```
 3.       Mars      227.9      3398      none        2   6.42e+23
 4.      Venus      108.2      6050      none        0   4.87e+24
 5.      Earth      149.6      6378      none        1   5.98e+24
 6.    Neptune     4496.6     24750     rings        8   1.03e+26
 7.     Uranus     2869.6     26145     rings       15   8.66e+25
 8.     Saturn       1427     60000     rings       17   5.69e+26
 9.    Jupiter      778.3     71900     rings       16   1.90e+27
```

The planets were originally entered in distance-from-sun order, so we could regain the original sequence by:

```
. sort dsun
```

In other datasets, we might want to create case ID numbers soon after the data are first entered:

```
. generate id = _n
. label variable id "Case ID number"
. save, replace
```

This ensures that no matter how much **sort**ing we subsequently do, we can always return to the original order of cases by typing

```
. sort id
```

**sort** can apply to several variables at once. Recalling the Jamaican wind power example from Chapter 1:

```
. use c:\stustata\wind, clear
(Jamaica Wind (Chen et al. 1990))
. describe
```

```
Contains data from c:\stustata\wind.dta
  Obs:    22 (max=  2620)                Jamaica Wind (Chen et al. 1990)
  Vars:    4 (max=    99)
 Width:   20 (max=   200)
    1. site       str15  %15s              Station
    2. side       byte   %8.0g      slbl   Which Side of Island?
    3. elev       int    %8.0g             Elevation in Meters
    4. wind       int    %8.0g             Avg. Wind Power/meter^2
Sorted by:
```

By typing **sort side wind**, we sort the cases first by side of island, then for each side, by wind power:

```
. sort side wind
. list site side wind
```

```
                 site      side     wind
  1.      Fullerswood     south       14
  2.           Bodles     south       31
  3.         Crawford     south       34
  4.    Rowlandsfield     south       55
  5.          Yallahs     south      100
  6.         Hillside     south      140
  7.           Manley     south      188
  8.     Morant Point     south      193
  9.        Hellshire     south      198
 10.        Spur Tree     south      199
 11.         Flagaman     south      200
 12.            Munro     south      237
 13.      Mason River     north       24
 14.       Manchoneal     north       50
 15.  Passley Gardens     north       70
 16.           Vinery     north       70
 17.       Folly Point     north       93
```

```
18.     Fairy Hill      north     134
19.      Sangster       north     145
20.    Pimento Hill     north     200
21.   Discovery Bay     north     200
22.       Galina        north     236
```

*side* is actually a labeled numeric variable {0="south", 1="north"}, which explains why **sort side** puts "south" before "north." If *side* were a string variable instead, **sort side** would put "north" and "south" in alphabetical order.

## From a Raw-Data File

**input** provides a fast way to type in small datasets. With larger projects, it is often more efficient to initially type and edit data with a word processor, spreadsheet, or database. Table 2 lists data from Brown et al. 1986:

**Table 2: Estimated Decommissioning Costs for Five Nuclear Power Plants**

| Site | Capacity (megawatts) | Decommissioning Costs (million $) | Year Operation Started | Year Operation Closed |
|------|---------------------|-----------------------------------|------------------------|-----------------------|
| Elk River | 24 | 14 | 1962 | 1968 |
| Windscale | 33 | 64 | 1963 | 1981 |
| Humboldt Bay 3 | 65 | 55 | 1963 | 1976 |
| Shippingport | 72 | 98 | 1957 | 1982 |
| Dresden 1 | 210 | 95 | 1960 | 1978 |

We could start by typing these data directly into our word processor, saving it as an ASCII (text) file named *reactor.raw*:

```
"Elk River" 24 14 1962 1968
Windscale 33 64 1963 1981
"Humboldt Bay 3" 65 55 1963 1976
Shippingport 72 98 1957 1982
"Dresden 1" 210 95 1960 1978
```

Enclose strings like the reactor site names in quotations if they include blank spaces. Separate numerical variable values by at least one space. It does not matter whether the columns line up, or a case requires more than one line. If missing values exist, do not leave them as blanks—type in a period, preceded and followed by at least one space. These rules are the same mentioned earlier regarding **input**.

**infile** has a syntax similar to **input**, but it reads raw data from a text file rather than from the keyboard:

```
. infile str30 site capacity decom start close using c:\stustata\reactor.raw
(5 observations read)

. list
              site    capacity      decom      start      close
   1.     Elk River        24         14       1962       1968
   2.     Windscale        33         64       1963       1981
   3.  Humboldt Bay 3      65         55       1963       1976
   4.   Shippingport       72         98       1957       1982
   5.     Dresden 1       210         95       1960       1978

. label data "Reactor Decommissioning Costs"
```

```
. label variable site "Reactor site"
. label variable capacity "Capacity in megawatts"
. label variable decom "Decommissioning cost, $millions"
. label variable start "Year operations started"
. label variable close "Year operations closed"
. compress
```

```
site was str30 now str14
capacity was float now int
decom was float now byte
start was float now int
close was float now int
```

```
. save c:\stustata\reactor
```

File *reactor.dta* is now labeled, compressed, and saved in Stata format. Note that **compress** frees us from having to know the exact length of our string variables. Just specify more than enough width (as we did with **infile str30** *site*), and let **compress** pare it down to the minimum needed. Here, **str14** turns out to be enough, since no site names have more than 14 characters.

## From Other Programs

A commercial program called Stat/Transfer, manufactured by Circle Systems (Seattle, WA), performs easy translation back and forth among the following dataset formats:

| Database | Spreadsheet | Statistical |
|---|---|---|
| Alpha Four | Excel | Gauss |
| Clipper | Lotus 1-2-3 (all) | SAS Transport |
| dBase II, III, or IV | Quattro Pro | SPSS Export |
| Foxbase | Symphony | Stata |
| Paradox | | Systat |

For example, Stat/Transfer can automatically build a Stata dataset from an SPSS export file or vice versa.

Another program, DBMS/COPY, provides similar capabilities for translation among a larger list of program formats. See Hilbe (1991, 1992) for a comparative review of these two programs, from the viewpoint of a Stata user. Among his conclusions:

1.  If you download files from mainframe SPSS or SAS, Stat/Transfer appears ideal.

2.  Stat/Transfer is adequate if you use Lotus 1-2-3 or dBase compatible programs.

3.  If you work strictly within the PC domain and use a variety of statistical, spreadsheet, and database programs, DBMS/COPY may be more valuable.

4.  Stat/Transfer is less expensive.

These programs are available from:

| | |
|---|---|
| DBMS/COPY | Stat/Transfer |
| SPSS Inc. | Computing Resource Center |
| 444 N. Michigan Ave. | 1640 Fifth Street |
| Chicago, IL 60611 | Santa Monica, CA 90401 |
| telephone:    800 543-2185 | telephone:    800 782-8272 |
| | Fax:              310 393-7551 |

For large projects, translation programs like Stat/Transfer or DBMS/COPY permit a very efficient data-entry strategy. Use your favorite database to originally enter and store the data, perhaps setting up and typing into an onscreen data-entry form. When data entry is complete, use the translation program to move it to Stata format. Sometimes an intermediate step is necessary. For example, I use Reflex database and Stat/Transfer, which does not support Reflex. But Reflex can save files in Paradox format, and Stat/Transfer translates Paradox format to Stata. Even this two-step translation process is nearly effortless, taking at most a few minutes.

## Combining Datasets

Stata can combine datasets in two different ways:

**append**  The two datasets involve different sets of cases.  Suppose *oldfile1.dta* contains information on variables *age* , *income* , and *education* for cases 1 through 92, and *oldfile2.dta* contains information on *age*, *income*, and *education* for cases 93 through 136.  To combine them:

```
. use oldfile1
. append using oldfile2
. save newfile
```

**append** "lengthens" the dataset, like taping the top of a new sheet of paper to the bottom of your old sheet.  If one dataset contains some variables not included in the other, Stata fills in with missing values as needed.

**merge**  The two Stata-format datasets have (at least some of) the same cases, but different sets of variables.  Suppose *oldfile.dta* contains *age* and *income*, plus case identification numbers *id*.  A separate dataset called *morevars.dta* records political party (*party*), presidential candidate preference (*prez*), and *id* numbers for many of these same voters.  To match observations having the same *id* numbers, and combine *age*, *income*, *party*, and *prez* into one new dataset called *newfile.dta*, we first assure that both datasets are sorted in *id* order:

```
. use morevars
. sort id
. save, replace
. use oldfile
. sort id
. merge id using morevars
. save newfile
```

**merge** "widens" the dataset, somewhat like taping two sheets of paper together side by side.  In the example above, new dataset *newfile.dta* will contain variables *id*, *age*, *income*, *party*, *prez*, and an automatically defined variable called _*merge*. _*merge* indicates the original source of information:

_*merge* = 1     The case occurred only in the master dataset (e.g., *oldfile*).
_*merge* = 2     The case occurred only in the **using** dataset (e.g., *morevars*).
_*merge* = 3     The case occurred in both datasets.

Type **help merge** for more examples using this command.

## Large Datasets (Professional Stata Only)

Professional Stata can handle up to 254 variables and 32,000 cases (many more using Intercooled Stata, OS/2, or Unix), if enough memory is available. Typing **describe** at any time reveals the current limit on variables and cases:

```
. describe
```

```
Contains data
  Obs:       0  (max=   2620)
  Vars:      0  (max=     99)
Width:       0  (max=    200)
Sorted by:
```

Stata shows room for 2620 cases and 99 variables, given the memory (only 512K) presently allocated to Stata on my computer. (With 4M memory, DOS Stata would show 20,540 cases.) An error message appears if we try to use a dataset larger than the maximums listed. For example, if I have a file called *records.dta* containing 3000 cases and 20 variables:

```
. use records
no room to add more observations
r(901);
```

I can reallocate memory, trading off fewer variables for more cases, with **set maxobs**:

```
. set maxobs 3100
. describe
```

```
Contains data
  Obs:       0  (max=   3119)
  Vars:      0  (max=     83)
Width:       0  (max=    168)
Sorted by:
```

I could now use a 3000 by 20 file like *records.dta*.

To make room for a file with fewer cases but more variables, increase **maxvar**:

```
. set maxvar 254 width 600
. describe
```

```
Contains data
  Obs:       0  (max=    869)
  Vars:      0  (max=    254)
Width:       0  (max=    602)
Sorted by:
```

When we increase **maxvar**, Stata automatically compensates by reducing **maxobs**. The **width 600** part of this command is optional, but it ensures that we really can load 254 variables (see **help maxvar** or **help memsize**).

On DOS computers, Professional Stata can recognize and use expanded memory. Intercooled Stata, which requires a 386/387 SX or DX, 486DX, or 586 processor, uses extended memory. Older computers, with only conventional DOS memory (640K) available, become a cramped environment for working with datasets of more than about 6000 cases or 100 float variables (or about 200 int or 400 byte variables). Even a small amount of expanded memory will bring noticeable improvements.

A shortcut for **infile**-ing large datasets is to give all the variables similar, numbered names. For example, if we have data on 97 variables in file *bigfile.txt*:

```
. infile var1-var97 using bigfile.txt
```

We can rename variables later, if desired. For example:

```
. rename var1 age
. rename var2 income
```

`infile` and labeling commands get awkward and error-prone when we have many variables. You may find it easier to `infile` and `label` large datasets by writing a simple program called a do-file (next section).

## Stata Programming

Any sequence of valid Stata commands, typed on a word processor and saved as a text file with a `.do` extension, can serve as a simple Stata program. For example, here is a do-file that automatically performs the same steps taken earlier to create a Stata-format dataset on nuclear power plants. It reads in a raw data file called *reactor.raw*, labels data and variables, compresses, and saves:

```
clear
infile str14 site capacity decom start close using c:\stustata\reactor.raw
label data "Reactor Decommissioning Costs"
label variable site "Reactor site"
label variable capacity "Capacity in megawatts"
label variable decom "Decommissioning cost, $millions"
label variable start "Year operations started"
label variable close "Year operations closed"
compress
save c:\stustata\reactor, replace
```

After typing these lines with your word processor, save them as text file *makedata.do*. Be sure to type them exactly as shown (you may need to set wide margins), and hit <Enter> after typing the last line. If you saved *makedata.do* in directory *C:\stustata*, then typing

. **do** *c:\stustata\makedata.do*

creates the dataset.

A do-file can be run any number of times. This makes it easy to update a large dataset as more data come in. Simply type new data into the original raw-data file, and rerun the do-file. For example, at some future time we might add more nuclear power plants to *reactor.raw*. Again typing

. **do** *c:\stustata\makedata.do*

will construct a new, complete dataset.

Instead of simply executing a sequence of commands, do-files can also be used to define a program that will stay available in memory for the duration of a Stata session. To do this, begin the do-file with **program define** *progname* and end it with **end**. For example, this do-file defines a new command called **tcorr**, that performs a *t* test of the Pearson correlation between two variables (something that Stata itself does not routinely do):

```
program define tcorr
    macro define _Y "%_1"
    macro define _X "%_2"
    correlate %_Y %_X
    quietly regress %_Y %_X
    display " t = "_b[%_X]/_se[%_X]
    display "Prob > |t| = "tprob(_result(5),_b[%_X]/_se[%_X])
end
```

Though quite primitive, *tcorr.do* illustrates several basic Stata programming tricks such as macros, quietly performed (no output) analysis, and recalling of specific results from the previous analysis. `_b[`*varname*`]` and `_se[`*varname*`]` represent the regression coefficient on *varname* and its standard error, from the previous regression. `_result(5)` is that regression's residual degrees of freedom.

To utilize *tcorr.do*, first type the program on your word processor and save it as a text file. Next get into Stata, and run *tcorr.do* to load it into memory:

```
. run c:\stustata\tcorr.do
```

From then on, the newly defined **tcorr** command should work anytime you want during the remainder of the session. To test the correlation between Jamaican wind and elevation, for example:

```
. use c:\stustata\wind, clear
(Jamaica Wind (Chen et al. 1990))
. tcorr wind elev
(obs=21)

        |    wind      elev
--------+------------------
   wind|  1.0000
   elev|  0.2563   1.0000

 t = 1.1557296
Prob > |t| = .26211935
```

A do-file named with an `.ado` extension becomes an ado-file, or automatic do-file. With ado-files, we do not need to first type a **do** command. Many Stata procedures actually are ado-files. On DOS and OS/2 computers, Stata looks for ado-files in directories *C:\stata\ado* (*C:\stustata\ado* if using Student Stata) and *C:\ado*. If we type, at the start of a Stata session, something like

```
. macro define S_ADO "d:\somewher\else"
```

then Stata thereafter looks for ado-files in the *D:\somewher\else* directory.

Both do- and ado-files employ Stata's programming language, which provides users with tools to substantially extend Stata's native capabilities. Other examples in this book appear on pages 120–121, 176–177, and 183–184. To see more polished programming work, study some of the hundreds of ado-files supplied with Stata. The *Stata Reference Manual*, **help ado**, **help do**, and **help program** all supply further details.

## Generating New Variables

The **generate** command creates new variables from algebraic expressions. Returning to *reactor.dta*:

```
. use c:\stustata\reactor
. describe
```

```
Contains data from c:\stustata\reactor.dta
  Obs:     5 (max=  2620)              Reactor Decommissioning Costs
  Vars:    5 (max=    99)
Width:    21 (max=   200)
   1. site         str14   %14s       Reactor site
   2. capacity     int     %8.0g      Capacity in megawatts
   3. decom        byte    %8.0g      Decommissioning cost, $millions
   4. start        int     %8.0g      Year operations started
   5. close        int     %8.0g      Year operations closed
Sorted by:
```

We could generate a new variable called *years*, the number of years a reactor stayed in operation:

```
. generate years = close - start
. label variable years "Years in operation"
```

Similarly, we might want a variable giving the decommissioning cost per year of operation:

```
. generate peryear = decom/years
. label variable peryear "Decom. cost/years operated"
. save, replace
file c:\stustata\reactor.dta saved
```

File *reactor.dta* now contains two new variables:

```
. describe
```

```
Contains data from c:\stustata\reactor.dta
  Obs:       5 (max=  2620)                Reactor Decommissioning Costs
  Vars:      7 (max=    99)
Width:      29 (max=   200)
    1. site            str14   %14s        Reactor site
    2. capacity        int     %8.0g       Capacity in megawatts
    3. decom           byte    %8.0g       Decommissioning cost, $millions
    4. start           int     %8.0g       Year operations started
    5. close           int     %8.0g       Year operations closed
    6. years           float   %9.0g       Years in operation
    7. peryear         float   %9.0g       Decom. cost/years operated
Sorted by:
```

```
. list site start close years decom peryear
```

|    |           site | start | close | years | decom | peryear |
|----|---------------|-------|-------|-------|-------|---------|
| 1. |     Elk River | 1962  | 1968  | 6     | 14    | 2.333333 |
| 2. |     Windscale | 1963  | 1981  | 18    | 64    | 3.555556 |
| 3. | Humboldt Bay 3 | 1963  | 1976  | 13    | 55    | 4.230769 |
| 4. |  Shippingport | 1957  | 1982  | 25    | 98    | 3.92    |
| 5. |     Dresden 1 | 1960  | 1978  | 18    | 95    | 5.277778 |

**generate** can create new variables from any mixture of old variables, constants, and algebraic expressions. It recognizes the following algebraic symbols:

| | |
|---|---|
| + | addition |
| – | subtraction |
| * | multiplication |
| / | division |
| ^ | raise to power |

Parentheses will control the order of calculation; without them, the ordinary rules of precedence apply ($1+3*4 = 13$, but $(1+3)*4 = 16$). Stata also recognizes these mathematical functions:

| | |
|---|---|
| abs(x) | absolute value |
| atan(x) | arctangent returning radians |
| cos(x) | cosine of radians |
| exp(x) | exponent ($e$ to power) |
| gammap(a,c) | incomplete gamma $P(a, x)$ |
| ibeta(a,b,x) | incomplete beta $I_x(a, b)$ |
| int(x) | integer from truncating $x$ |
| ln(x) | natural (base $e$) logarithm |
| lngamma(x) | $\ln(\Gamma(x))$ |
| log(x) | same as ln(x) |
| mod(x,y) | modulus $x$ with respect to $y$ |
| round(x,y) | rounds $x$ into units of $y$ |
| sin(x) | sine of radians |
| sqrt(x) | square root |

Other mathematical functions can be derived from Stata's built-in functions. For example:

base 10 logarithms:    take natural logs and divide by the natural log of 10. To **generate** a new variable, *logdecom*, equal to the base 10 log of *decom*:

. generate *logdecom* = ln(*decom*)/ln(10)

factorials:    to obtain $x$ factorial ($x!$),

. generate *xfact* = round(exp(lngamma(*x*+1)),1)

$\ln(x!)$ equals lngamma($x$+1); logarithms of factorials may be easier to work with than the factorials themselves when numbers are large.

sine or cosine    Stata's trigonometric functions (**sin**, **cos**, and **atan**) are defined
of degrees:    in terms of radians. Degrees may be converted to radians by the formula $r = d\pi/180$, where $d$ represents degrees and $r$ represents radians. If our data contain a variable called *angle* in degrees, and we want the sine of this angle:

. generate *sinangle* = sin(*angle*\*_pi/180)

The expression in parentheses converts degrees to radians. _pi is Stata's symbol for the mathematical constant $\pi$.

Other trigonometric functions, assuming $x$ is measured in radians are:

tangent of $x$:    . generate *tanx* = sin(*x*)/cos(*x*)

arcsine ($\sin^{-1}$) of $x$:    . generate *arcsinx* = atan(*x*/sqrt(1-*x*^2))

arccosine ($\cos^{-1}$) of $x$:    . generate *arccosx* = atan(sqrt(1-*x*^2)/*x*)

A progression of mathematical re-expressions called the "Ladder of Powers" often helps in solving statistical problems such as outliers, nonnormality, and nonlinearity.

**Table 3: Ladder of Powers**

| Transformation | Stata | Effect |
|---|---|---|
| cube: $Y^3$ | *newY* = *oldY*^3 | reduce severe negative skew |
| square: $Y^2$ | *newY* = *oldY*^2 | reduce mild negative skew |
| raw: $Y^1$ | *oldY* | raw data |
| square root: $Y^{1/2}$ | *newY* = *oldY*^.5 | reduce mild positive skew |
| logs: $\log_e(Y)$<br>or $\log_{10}(Y)$ | *newY* = ln(*oldY*)<br>*newY* = ln(*oldY*)/ln(10) | reduce positive skew |
| negative reciprocal<br>root: $-(Y^{-1/2})$ | *newY* = -(*oldY*^-.5) | reduce severe positive skew |
| negative reciprocal:<br>$-(Y^{-1})$ | *newY* = -(*oldY*^-1) | reduce very severe positive skew |

For example, if we find that variable *distance* exhibits positive skew, we might prefer to work with logarithms of *distance*:

. generate *logdist* = ln(*distance*)

or, if *distance* includes some zero values,

```
. generate logdist = ln(distance+1)
```

Unlike categorization or some other transformation schemes, power transformations are reversible. That is, we can always recover the original values exactly by applying an appropriate inverse transformation (Table 4).

**Table 4:  Inverse Transformations for Ladder of Powers**

| Ladder of Powers Transformation | Inverse Transformation | Stata Inverse Transformation |
|---|---|---|
| cube:  $Y^3$ | $Y^{1/3}$ | `newY^(1/3)` |
| square:  $Y^2$ | $Y^{1/2}$ | `newY^(1/2)` |
| raw:  $Y^1$ | no change | |
| square root:  $Y^{1/2}$ | $Y^2$ | `newY^2` |
| logs: $\log_e(Y)$ or:  $\log_{10}(Y)$ | $e^Y$ $10^Y$ | `exp(newY)` `10^newY` |
| negative reciprocal root:  $-(Y^{-1/2})$ | $(-Y)^{-2}$ | `-newY^-2` |
| negative reciprocal: $-(Y^{-1})$ | $(-Y)^{-1}$ | `-newY^-1` |

Reversibility proves that no information is lost in transformation.  Chapters 6 and 9 illustrate one important use for inverse transformations (curvilinear regression).

Stata has eight built-in statistical functions:

**Binomial(n,k,π)**   probability of observing $k$ or more successes in $n$ trials when the probability of a single success is $\pi$

**chiprob(df,x)**   probability of a $\chi^2$ (chi-square) value of $x$ or larger, $P(\chi^2 \geq x)$, given a $\chi^2$ distribution with $df$ degrees of freedom

**fprob(df1,df2,f)**   probability of an $F$ value of $f$ or larger, $P(F \geq f)$, given an $F$ distribution with $df1$ and $df2$ degrees of freedom

**invbinomial(n,k,p)**   inverse binomial: for $p < .5$, returns probability $\pi$ such that the probability of observing $k$ or more successes in $n$ trials is $p$; for $p > .5$, returns probability $\pi$ such that the probability of observing $k$ or fewer successes in $n$ trials is $1-p$

**invnorm(p)**   $z$ such that the probability of an equal or lower value is $p$, $P(Z \leq z) = p$, given a standard normal distribution

**invt(df,p)**   $t$ such that the probability of an equal or lower absolute value is $p$, $P(|T| \leq t) = p$, given a $t$-distribution with $df$ degrees of freedom

**normprob(z)**   probability of a value of $z$ or larger, $P(Z \geq z)$, given a standard normal distribution

**tprob(df,t)**   probability that the absolute value of $T$ exceeds $t$, $P(|T| \geq t)$, given a $t$ distribution with $df$ degrees of freedom

All of the algebraic, mathematical, and statistical functions described in this section work either with **generate** (create a new variable) or **replace** (change values of an old variable). A third place to employ them is with Stata's hand-calculator function, `display`. For example, if we want to find out what 3 + 4 equals:

```
. display 3+4
7
```

or, making a less obvious calculation:

```
. display ln(5280)/sqrt(0.92)
8.9365956
```

Unlike `generate` or `replace`, the `display` command has no effect on the data in memory. It merely displays the answer to the problem we type in.

Further information on some related topics can be obtained by typing `help functions` (a complete list of `generate`, `replace`, and `display` functions, including string and other special functions). Still more functions, including counts, moving averages, medians, ranks, and standardized values plus many others, are available through the `egen` (extended `generate`) command. Typing `help egen` obtains this command's full syntax and other details. For a formal approach to normalizing transformations, see `help boxcox` (maximum-likelihood Box-Cox transformation).

## Random Numbers and Random Sampling

Stata can produce pseudo-random numbers. Since they result from a mathematical equation, such numbers are not truly random, but for most practical purposes they appear to be random. This equation begins with a "seed value," used to generate the first number, then uses that number in turn as a seed to generate the second, and so on. Unless we tell it otherwise, Stata generates its first random number *for any session* with the seed value 1001. Each subsequent number then uses the previous result, for as many random numbers as we need. Beginning another session with the same default seed value will produce the same sequence of pseudo-random numbers. If you do not want this to happen, use the `set seed` command to specify a different starting seed. Any large, positive odd number will do. For example (using my daughter's birthday):

```
. set seed 31977
```

Stata uses this seed as the first value of $R_n$, to find:

$$R_{n+1} = 69069R_n \qquad (\text{mod } 2^{32})$$

The first random number generated is the fractional part of $R_{n+1}/2^{31}$.

$R_{n+1}$ then becomes the next seed:

$$R_{n+2} = 69069R_{n+1} \qquad (\text{mod } 2^{32})$$

so the second random number is the fractional part of $R_{n+2}/2^{31}$, and so on for as many random numbers as needed. With an accurate calculator you may work out for yourself why Stata's first two random numbers, following `set seed 31977`, are .0284686 and .2949037.

The `uniform()` function generates random numbers:

```
. generate randnum = uniform()
```

New variable *randnum* will consist of apparently random 16-digit numbers between 0 and 1. Every 16-digit number between 0 and 1 has the same probability of occurring, so this is called a "uniform probability distribution."

A simple application for random numbers is drawing random samples from our data. To illustrate, we turn again to the Jamaican winds data, and **generate** a new variable containing random numbers from a uniform probability distribution:

```
. use c:\stustata\wind
(Jamaica Wind (Chen et al. 1990))
. generate rand = uniform()
. list
```

|     | elev | wind | site | rand |
|-----|------|------|------|------|
| 1.  | 30   | 31   | Bodles | .0321949 |
| 2.  | 30   | 34   | Crawford | .6710823 |
| 3.  | 0    | 200  | Discovery Bay | .9830985 |
| 4.  | 76   | 134  | Fairy Hill | .6296493 |
| 5.  | 60   | 200  | Flagaman | .2459989 |
| 6.  | 0    | 93   | Folly Point | .8984687 |
| 7.  | 0    | 14   | Fullerswood | .3359111 |
| 8.  | 0    | 236  | Galina | .0407609 |
| 9.  | 0    | 198  | Hellshire | .3131475 |
| 10. | 30   | 140  | Hillside | .781247 |
| 11. | 6    | 50   | Manchoneal | .9458091 |
| 12. | 0    | 188  | Manley | .0863035 |
| 13. | .    | 24   | Mason River | .8953475 |
| 14. | 0    | 193  | Morant Point | .754981 |
| 15. | 792  | 237  | Munro | .7855499 |
| 16. | 15   | 70   | Passley Gardens | .1429575 |
| 17. | 305  | 200  | Pimento Hill | .9293609 |
| 18. | 305  | 55   | Rowlandsfield | .0243085 |
| 19. | 0    | 145  | Sangster | .964518 |
| 20. | 610  | 199  | Spur Tree | .2912911 |
| 21. | 457  | 70   | Vinery | .1880395 |
| 22. | 60   | 100  | Yallahs | .7001474 |

To choose a random sample of five cases from these data, sort the data in random order and pick the first five:

```
. sort rand
. list in 1/5
```

|     | elev | wind | site | rand |
|-----|------|------|------|------|
| 1.  | 305  | 55   | Rowlandsfield | .0243085 |
| 2.  | 30   | 31   | Bodles | .0321949 |
| 3.  | 0    | 236  | Galina | .0407609 |
| 4.  | 0    | 188  | Manley | .0863035 |
| 5.  | 15   | 70   | Passley Gardens | .1429575 |

The first 10 cases (**list in 1/10**) would comprise a random sample of size 10, and so on.

Instead of a sample of specified size, we might want a sample of specified probability. To get an approximately 25% random sample, pick cases for which *rand* < .25. We could mark cases chosen for this sample by generating a logical variable called *sample25*, equal to 1 if chosen and zero otherwise:

```
. generate sample25 = rand < .25
. list
```

|     | elev | wind | site | rand | sample25 |
|-----|------|------|------|------|----------|
| 1.  | 305  | 55   | Rowlandsfield | .0243085 | 1 |
| 2.  | 30   | 31   | Bodles | .0321949 | 1 |
| 3.  | 0    | 236  | Galina | .0407609 | 1 |
| 4.  | 0    | 188  | Manley | .0863035 | 1 |
| 5.  | 15   | 70   | Passley Gardens | .1429575 | 1 |
| 6.  | 457  | 70   | Vinery | .1880395 | 1 |
| 7.  | 60   | 200  | Flagaman | .2459989 | 1 |
| 8.  | 610  | 199  | Spur Tree | .2912911 | 0 |

| 9. | 0 | 198 | Hellshire | .3131475 | 0 |
| 10. | 0 | 14 | Fullerswood | .3359111 | 0 |
| 11. | 76 | 134 | Fairy Hill | .6296493 | 0 |
| 12. | 30 | 34 | Crawford | .6710823 | 0 |
| 13. | 60 | 100 | Yallahs | .7001474 | 0 |
| 14. | 0 | 193 | Morant Point | .754981 | 0 |
| 15. | 30 | 140 | Hillside | .781247 | 0 |
| 16. | 792 | 237 | Munro | .7855499 | 0 |
| 17. | . | 24 | Mason River | .8953475 | 0 |
| 18. | 0 | 93 | Folly Point | .8984687 | 0 |
| 19. | 305 | 200 | Pimento Hill | .9293609 | 0 |
| 20. | 6 | 50 | Manchoneal | .9458091 | 0 |
| 21. | 0 | 145 | Sangster | .964518 | 0 |
| 22. | 0 | 200 | Discovery Bay | .9830985 | 0 |

Similarly, the command

```
. generate sample05 = rand < .05
```

would mark cases for an approximately 5% random sample, and so on. The **sample** command provides an alternative, direct way to select a random sample (and discard the rest) of the data in memory.

## Case Weights

Stata understands four types of case weighting:

- **aweight**  analytical weights, used in weighted least squares (WLS) regression and similar procedures
- **fweight**  frequency weights, representing the number of duplicated observations
- **pweight**  sampling weights, equal to the inverse of the probability that an observation is included due to sampling strategy
- **iweight**  importance weights, however you define "importance"

Most Stata analytical procedures permit one or more type of weighting. The weights themselves can be any suitable variable in the data. Chapters 4 and 6 illustrate some possibilities.

## Also Type help

- **ado**  ado-file (automatic do-file) manipulation
- **append**  combine two datasets with different cases
- **boxcox**  maximum-likelihood Box-Cox transformation towards normality
- **compress**  change variable types to most efficient
- **decode**  create string variable from value-labeled numeric variable
- **display**  "hand calculator" function
- **do**  execute commands from a file
- **egen**  extend **generate** with many additional statistical functions
- **encode**  create value-labeled numeric variable from string variable
- **format**  variable display format (decimal place, rounding, width)
- **functions**  complete list of mathematical, statistical, string and special functions
- **generate**  calculate new variable from algebraic expression
- **infile**  read raw-data (ASCII) file
- **input**  type in raw data
- **ladder**  search for "best" normalizing transformation
- **maxobs**  change allowable number of cases

| | |
|---|---|
| maxvar | change allowable number of variables |
| memsize | memory and dataset size limitations |
| merge | combine two datasets with different variables |
| order | reorder *variables* (not cases) in a dataset |
| outfile | create an ASCII data file as output |
| program | define and manipulate programs |
| replace | change variable values using algebraic expression |
| sample | select random sample, dropping cases not selected |
| save | save Stata dataset |
| set | set program parameters |
| sort | re-order cases according to values of listed variable(s) |
| string | non-numerical variables (names, text, etc.) |
| use | retrieve previously-**save**d Stata dataset |
| weight | case weighting (analytical, frequency, sampling, or importance) |

# 3
# Graphs

Stata's basic **graph** command can produce many different displays. The details of graph style and appearance depend on options specified after this command. The main styles are:

| | |
|---|---|
| **hist** | histogram |
| **box** | boxplot |
| **oneway** | one-dimensional scatterplot |
| **twoway** | two-dimensional scatterplot |
| **matrix** | scatterplot matrix |
| **bar** | bar chart |
| **pie** | pie chart |
| **star** | star chart |

Stata ado-files produce other, more specialized types of graphs. The usual Stata qualifiers—**in**, **if**, and **by**—work as expected with graphs. Each style also offers options to control what is to be graphed and how the image looks.

The on-screen graphics capabilities of Student and Professional Stata are virtually the same, but Student Stata cannot save or print graphs. With Professional Stata, on the other hand, we can save graphs to disk, print high-resolution copies, or input the graphs into other programs for editing, document writing, and so forth.

## Histograms

Histograms provide basic graphical views of measurement variable distributions. For example, *city.dta* contains data on the percentage of families below the poverty level in 146 U.S. cities:

```
. use c:\stustata\city
(Data on 146 U.S. Cities)

. describe

Contains data from c:\stustata\city.dta
  Obs:    146 (max=  2620)                    Data on 146 U.S. Cities
  Vars:     8 (max=    99)
 Width:    44 (max=   200)
    1. divorce      float    %9.2f            Divorces/1000 ages 15-59
    2. density      float    %9.0g            Population per square mile
    3. poor         float    %9.2f            Percent Families below Poverty
    4. city         float    %9.0g   citylbl  City
    5. victims      float    %9.2f            Homicide Victims/100k 80-84
    6. region       float    %9.0g   rlbl     Geographical Region
    7. black        float    %9.2f            Percent Population Black
    8. justify      float    %9.2f            Justifiable Homicide Ratio
Sorted by:
```

To see a simple histogram of the variable *poor* (Figure 3.1):

```
. graph poor, hist
```

If you are using Professional Stata, and wish to save a graph for subsequent printing, add a **saving(*drive:filename*)** option such as:

```
. graph poor, hist saving(a:figure1)
```

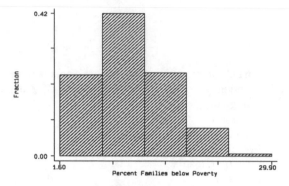

**Figure 3.1**

Figure 3.2 illustrates two ways to enhance a graph:

1.  **ylabel** and **xlabel** options tell Stata to use round-number labeling for the $Y$ and $X$ axes (compare with the default labeling seen in Figure 3.1).

2.  We can specify up to eight titles, two each at the top (**t1** and **t2**), bottom (**title** or **b1**, and **b2**), left side (**l1** and **l2**) and right side (**r1** and **r2**).

```
. graph poor, hist ylabel xlabel title(this is title)
        b2(this is title bottom 2) l1(this is title left 1)
        l2(this is title left 2) t1(this is title top 1) t2(this is title top
        2) r1(this is title right 1) r2(this is title right 2)
```

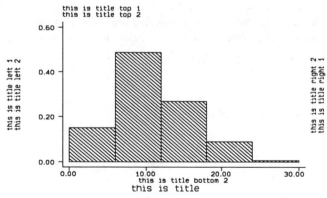

**Figure 3.2**

Some further options include

**bin()**     specifies the number of vertical bars; the default is **bin(5)**.

**noaxis**    suppresses the lines for $X$ and $Y$ axes.

**norm()**    draws a normal curve. **norm(10,3)**, for example, calls for mean 10, standard deviation 3. **norm** alone implies sample mean and standard deviation.

**ylabel()**  specifies what values should be labeled on the $Y$ axis. The default (if we just write **ylabel**) is to let Stata decide.

ytick()      specifies the location of *Y*-axis tick marks.

xlabel()     specifies what values should be labeled on the *X* axis.

xtick()      specifies the location of *X*-axis tick marks.

Figure 3.3 uses sample median and pseudo-standard deviation (*PSD*; see page 55) for normal-curve parameters:

```
. graph poor, hist bin(12) noaxis norm(10.25,5.33) ylabel(0,.1,.2)
     ytick(.05,.15) xlabel(0,10,20,30) xtick(5,15,25)
```

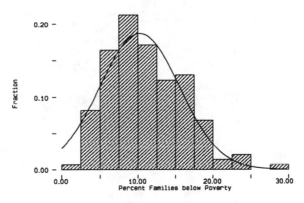

**Figure 3.3**

Suppose we want to compare the distribution of percent below poverty (*poor*) in different regions of the country (Figure 3.4):

```
. sort region
. graph poor, hist bin(9) norm by(region) ylabel xlabel
```

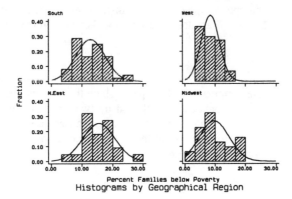

**Figure 3.4**

The **by** option obtains a separate histogram for each value of the categorical variable *region*. Note that the data first had to be sorted by *region*. To see a whole-sample histogram in addition to separate regional histograms, add the option **total** (Figure 3.5):

```
. graph poor, hist bin(9) norm by(region) ylabel xlabel total
```

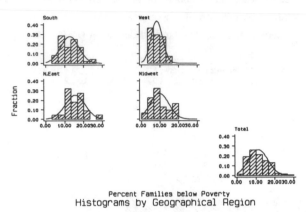

Percent Families below Poverty
Histograms by Geographical Region

**Figure 3.5**

Some general points about Stata's **graph** command:

1.  Its syntax is

    ```
    . graph variable-list, option1 option2 ...
    ```

    We can list whatever options we want in any order, so long as they come after the
    (single) comma.  To save graphs for printing or other use, include **saving(**drive:
    filename**)**  among the options (Professional Stata only).

2.  Except for **hist, bin()**, and **norm()**, the options illustrated in Figures 3.2–3.5
    work with other types of graphs besides histograms.  For example, **ylabel()**  also
    labels *Y*-axis values on scatterplots or boxplots.

3.  One way to get exactly the graph you want is to build it in steps:  start with the basic
    graph, view that, and add options one or two at a time.  Save only the final graph.  The
    <PgUp> key (<Ctrl><R>, Macintosh and Unix) makes this process easy.

## Time Plots

Stata does not have a separate command for time plots, since they are essentially just one kind
of scatterplot.  When a **graph** command is followed by two or more variable names, Stata
assumes we want a scatterplot unless told otherwise.  If the data form a time series and the last-
named variable is time, our scatterplot is also a time plot.  For example, to plot the worldwide
blue whale catch over 1920–1985 (Figure 3.6):

```
. use c:\stustata\whales
. describe

Contains data from c:\stustata\whales.dta
  Obs:     19 (max=  2620)                  World Whale Catch 1920-85
  Vars:     7 (max=    99)
Width:     28 (max=   200)
  1. year            float   %9.0g          Year
  2. blue            float   %9.0g          Blue Whales
  3. hump            float   %9.0g          Humpback Whales
  4. fin             float   %9.0g          Fin Whales
  5. sei             float   %9.0g          Sei Whales
```

```
   6. sperm          float    %9.0g                Sperm Whales
   7. total          float    %9.0g                All Whales
Sorted by:
```

`. graph blue year, connect(l) sort`

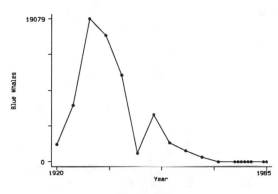

**Figure 3.6**

Seeing two variable names (*blue* and *year*), Stata assumes we want a scatterplot. The last-named variable (*year*) defines the $X$ axis of this scatterplot. Earlier-named variables (here only *blue*) form the $Y$ axis. To connect the plotted points with line segments, we include the **connect(l) sort** option. (Read carefully: that is a lower-case letter **l**, standing for "line segment," inside the parentheses of **connect(l)**.) The **sort** option, unlike a **sort** command, reorders the data for that one run only.

Stata allows up to twenty $Y$ variables, or twenty time series, per graph. Suppose we wanted a plot with two $Y$ variables (*blue* and *fin*), and better-labeled axes (Figure 3.7):

```
. graph blue fin year, connect(ll) sort ylabel(0,10000,20000,30000)
       xlabel(1920,1930,1940,1950,1960,1970,1980)
```

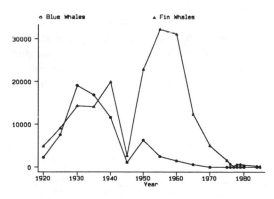

**Figure 3.7**

The **ylabel()** and **xlabel()** options work the same way here as with histograms.

**connect(ll)** asks Stata to connect both the first and the second-named $Y$ variables with line segments. We can control what symbols Stata uses to plot the points and how points are connected; see the Scatterplots and Stage sections of this chapter.

Stata's `plot` command produces simple text-mode time plots or scatterplots. `graph` is much more versatile, but `plot` offers the advantage of producing figures that can be written into a log file and saved or printed by Student Stata. For example, a text-mode plot corresponding to Figure 3.7:

`. plot blue fin year`

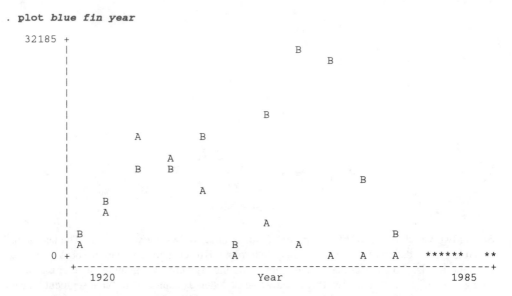

If we specify more than two variables, the first-named *Y* variable is plotted with A's, the second with B's, and so on. Asterisks mark locations where both variables overlap. Such text-mode plots do not look as nice as the images drawn by `graph`, but we can enhance them a bit using an old mainframe-researchers' trick: take a pen and connect the dots. Use different-color pens to connect the A's and B's.

## Scatterplots

Scatterplots have many other applications besides graphing time series (previous section). They are Stata's most versatile style of graph. A basic twoway scatterplot requires only `graph` followed by two variable names: `graph Y X`. For example, using American-city homicide rates (*victims*) and percent below poverty (*poor*), from *city.dta* (Figure 3.8):

`. graph victims poor`

We can add labels and titles as with other graph styles.

Scatterplots also permit weighting, where symbol sizes become proportional to the specified weighting variable. For example we could weight by cities' population densities (Figure 3.9):

```
. graph victims poor [iweight=density], ylabel(0,10,20,30,40,50)
       xlabel(0,10,20,30) title(Weighted Scatterplot)
```

The area of the circles in Figure 3.9 is proportional to density. `graph` permits analytic weights (`aweight`), frequency weights (`fweight`), and importance weights (`iweight`), treating all three types identically. Sampling weights (`pweight`) are not allowed.

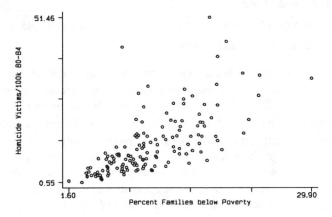

**Figure 3.8**

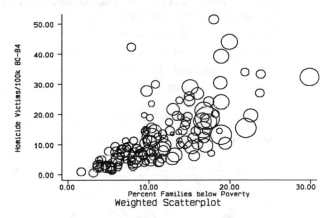

**Figure 3.9**

Scatterplot symbols can be any of the following:

| | |
|---|---|
| O | large circle |
| T | large triangle |
| S | large square |
| o | small circle |
| p | small plus sign |
| d | small diamond |
| . | dot |
| i | invisible |
| [_n] | case number |
| [*varname*] | value of variable *varname* |

For example, suppose we have variables named *Y1*, *Y2*, *Y3*, *X*, and *city*. To graph *Y1* versus *X*, using small diamonds:

```
. graph Y1 X, symbol(d)
```

To graph *Y1* (with small diamonds) and *Y2* (with large squares) against *X*:

```
. graph Y1 Y2 X, symbol(dS)
```

To graph *Y1* as diamonds, *Y2* as squares, and *Y3* using values of *city* as plotting symbols:

```
. graph Y1 Y2 Y3 X, symbol(dS[city])
```

Scatterplot points can be left unconnected, or connected any of these ways:

| | |
|---|---|
| **connect(l)** | Connect the points with line segments. Unless the data are already in order by *X*, you may want to specify **connect(l) sort**. |
| **connect(L)** | Connect with line segments so long as *X* keeps ascending. |
| **connect(m)** | Connect cross-medians for a series of vertical bands. The **bands()** option specifies how many bands; for example **connect(m) bands(6)** calls for dividing the data into six equal-width bands. |
| **connect(s)** | Connect cross-medians using smooth curves (cubic splines). As with **connect(m)**, we can also specify **bands()**. |
| **connect(.)** | Do not connect points; helpful if we have two or more *Y* variables. For example, **connect(.l)** does not connect the first *Y* variable, but connects the second with line segments. |
| **connect(J)** | Connect in steps (step function). |
| **connect(\|\|)** | Connect pairs of variables with vertical bars. |
| **connect(II)** | Same as **connect(\|\|)**, but bars capped like capital I's. |

**connect(\|\|)** and **connect(II)** are useful in high-low-close graphs and error-bar charts.

Plots of all possible pairs from a variable list appear in a scatterplot matrix (Figure 3.10):

```
. graph divorce density poor victims, matrix label symbol(p)
```

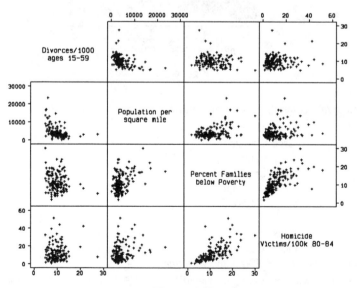

**Figure 3.10**

The **matrix** option calls for a scatterplot matrix, with variables in the order listed. **label** asks Stata to choose round-number labels for *Y-* and *X*-axis values. Another option, **half** (not used in Figure 3.10) suppresses printing of the redundant above-diagonal half of the matrix.

Twoway scatterplots can have boxplots or oneway scatterplots in their margins (Figure 3.11):

```
. graph victims poor, twoway oneway box title(Scatterplot with marginal
       boxplots and oneway scatterplots)
```

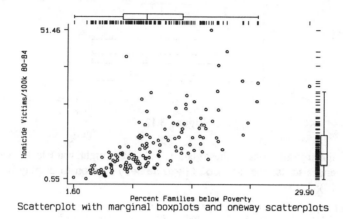

Scatterplot with marginal boxplots and oneway scatterplots

**Figure 3.11**

Figure 3.11 shows not only the joint distribution of *victims* and *poor* (central scatter), but also their individual, univariate distributions (margins). Marginal boxplots help keep us aware of univariate skewness or outliers. The following section describes other ways to use boxplots.

## Boxplots

Boxplots convey information about center, spread, symmetry, and outliers at a glance. To obtain a basic boxplot of *victims* (homicide rate) from dataset *city.dta* (Figure 3.12):

```
. graph victims, box ylabel
```

Labels, titles, and other embellishments could be added as usual. Figure 3.12 reveals a positively skewed distribution with outliers: cities with exceptionally high homicide rates.

The box in a boxplot extends from first to third quartiles, enclosing the interquartile range. Stata's boxplots approximate quartiles in the same manner as **summarize, detail**. This is not the same quartile approximation used in finding "fourths" for letter-values displays, **lv** (see Chapter 4). See Frigge, Hoaglin, and Iglewicz (1989) or Hamilton (1992b) for more about quartile approximations and their role in identifying outliers.

Like scatterplots, boxplots are versatile tools. Figure 3.11 illustrated how boxplots can be combined with oneway and/or twoway scatterplots. Another arrangement, parallel boxplots, allows quick visual comparison of several distributions at once. For example, we might compare U.S. regions in terms of their urban homicide rates (not shown):

```
. sort region
. graph victims, box by(region)
```

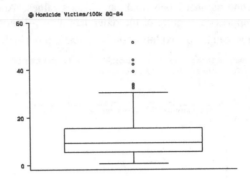

**Figure 3.12**

The overall median for all cities in these data is 9.2.  We might want to indicate this with a horizontal line, and also use the **symbol([varname])**  option to display names of outlier cities (Figure 3.13):

```
. graph victims, box by(region) yline(9.2)
         ylabel(0,10,20,30,40,50) symbol([city]) psize(130)
```

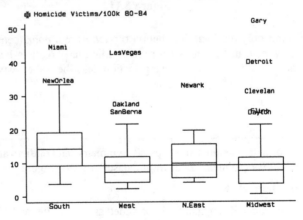

**Figure 3.13**

**symbol()**  affects only the outliers plotted in a boxplot.  The **psize(130)**  option enlarges plotting symbols (here, the city names) to 130% of their normal size.

Figure 3.13 shows parallel boxplots for the same variable (*victims*) across subgroups specified with the **by** option.  To add oneway scatterplots for each subgroup (not shown), type:

```
. graph victims, box oneway by(region)
```

We can also get parallel boxplots for several different variables, with commands of the form:

```
. graph var1 var2 var3, box
```

## Symmetry and Quantile Plots

Boxplots and histograms summarize distributions, hiding individual data points to simplify overall patterns. Symmetry and quantile plots, in contrast, include points for every case in the distribution. They are harder to read than summary graphs, but convey more detailed information.

A histogram of per-capita energy consumption in 128 countries (*energy.dta*) appears in Figure 3.14. The skewed distribution includes a handful of very high-consumption nations.

```
. graph energy, hist bin(9) norm
```

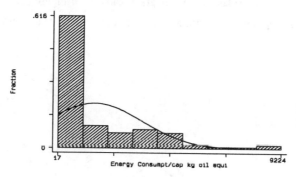

**Figure 3.14**

Figure 3.15 depicts this distribution as a symmetry plot. It plots the distance of the *i*th case above the median (vertical) against the distance of the *i*th case below the median. All points would lie on the diagonal line if this distribution were symmetrical. Instead we see that distances above the median grow steadily larger than corresponding distances below the median, symptomatic of positive skew. Unlike Figure 3.14, Figure 3.15 also reveals that the energy-consumption distribution is approximately symmetrical near its center.

```
. symplot energy
```

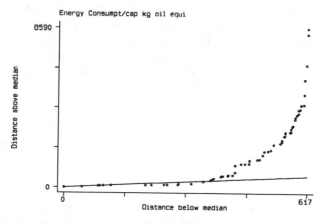

**Figure 3.15**

Quantiles are values below which a certain fraction of the data lie. For example, a .3 quantile is that value higher than 30% of the data. If we sort $n$ observations in ascending order, the $i$th value forms the $(i-.5)/n$ quantile. To calculate quantiles of the variable *energy*:

```
. drop if energy==.
. sort energy
. generate quantile = (_n-.5)/_N
```

(_n and _N are Stata system variables: _n represents the current case number, and _N the total number of cases.) Quantile plots automatically calculate what fraction of the cases lie below each actual data value, and display the results graphically as in Figure 3.16.

```
. quantile energy, ylabel(0,1000,2000,3000,4000,5000,6000,7000,8000,9000)
```

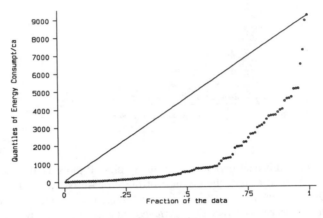

**Figure 3.16**

From well-labeled quantile plots, we can estimate order statistics, such as median (.5 quantile) or quartiles (.25 and .75 quantiles). *IQR* equals the rise between .25 and .75 quantiles. We can also read quantile plots to estimate the fraction of cases falling below a given value.

Quantile-quantile (*Q-Q*) plots compare two distributions at every point. Figure 3.17 graphs quantiles of energy consumption (vertical) against quantiles of gross national product. If the two distributions were identical, we would see points along the diagonal line. Instead, data points form a roughly straight line that is not parallel to the diagonal—indicating these two distributions have similar shapes, but different standard deviations and means.

```
. qqplot energy gnpcap
```

Quantile-quantile plots can also compare empirical distributions with theoretical distributions. For example, quantile-normal (*Q*-normal) plots graph quantiles of a variable's distribution against quantiles of a normal or Gaussian distribution (Figure 3.18).

```
. qnorm energy
```

If a distribution is approximately normal, quantile-normal plot points will lie along the diagonal line. Departures from this line provide detailed information about nonnormality, and also guide efforts to find normalizing transformations (for example, using the Ladder of Powers). *Regression with Graphics* includes an introduction to reading quantile-based plots. Chambers et al. (1983) provide more details.

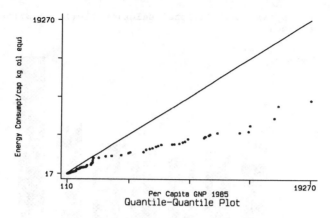

**Figure 3.17**

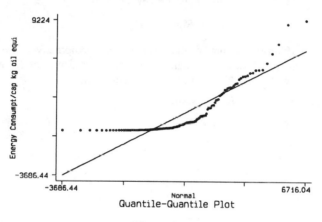

**Figure 3.18**

## Bar Charts, Pie Charts, and Star Charts

In their simplest application, bar charts display frequency distributions for categorical variables and look similar to histograms. They can convey more complex information as well. Dataset *xmasale.dta* contains information on a department store's daily sales receipts during seven weeks in November to January:

```
Contains data from c:\stustata\xmasale.dta
  Obs:    70 (max=  2620)            Dept. Store Christmas Sales
  Vars:    6 (max=    99)
Width:    10 (max=   200)
   1. date          int     %8.0g              Date:  13=January 1988
   2. weekday       byte    %8.0g       day     Day of the Week
   3. ad            byte    %8.0g       ad      Advertised Sale in Progress
   4. sales         float   %9.0g               Gross Sales
   5. temp          byte    %8.0g       temp    Temperature Category
   6. weather       byte    %8.0g       weather Weather Category
Sorted by:  date
```

Suppose we wished to see the total amount sold on each day of the week (Figure 3.19):

```
. sort weekday
. graph sales, bar by(weekday) ll(Total Sales) title(Total Sales
        by Day of Week)
```

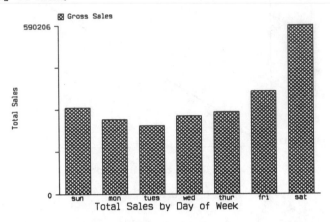

**Figure 3.19**

The bars in Figure 3.19 represent sums of the variable *sales*. Alternatively, we might prefer to show means. For example, how does weather affect sales? Mean sales for each category of weather are most easily obtained by the **tabulate** command:

```
. tabulate weather, summ(sales) means
```

```
    Weather| Summary of Gross Sales
   Category|        Mean
-----------+------------
      fair |    35450.882
      rain |      42335.5
      snow |    21103.571
    cloudy |        31181
-----------+------------
     Total |    34002.176
```

Bar charts display the same information graphically (Figure 3.20):

```
. sort weather
. graph sales, bar by(weather) means ll(Mean Sales) ylabel
        title(Mean Sales by Weather Category)
```

The **means** option produces bars that are proportional to the mean sales in each category of the **by**-group, *weather*. As before, a left-margin label was specified: **ll(Mean Sales)**.

Business reports often employ bar charts in which sums for different variables are stacked atop one another. This requires specifying more than one variable and adding the **stack** option. If we list more than one variable without **stack**, the bars appear side by side.

Pie charts are another very popular style of graph, used to show what proportion of a total belongs in each of several categories. For example, the data in *housproj.dta* describe 26 Boston-area family housing projects. Variables include the number of white, black, Hispanic, and "other" heads of households:

```
Contains data from c:\stustata\housproj.dta
  Obs:     26 (max=  2620)              Housing Projects (Boston Globe)
  Vars:     7 (max=    99)
Width:     11 (max=   200)
   1. id            byte   %8.0g
```

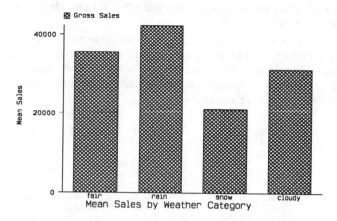

**Figure 3.20**

```
 2. project      byte    %8.0g    plbl    Project
 3. white        int     %8.0g            White
 4. black        int     %8.0g            Black
 5. hispanic     int     %8.0g            Hispanic
 6. other        byte    %8.0g            other
 7. total        int     %8.0g            total # household heads
Sorted by:
```

A pie chart can show the relative proportions of each ethnic group in all 26 housing projects combined (Figure 3.21):

```
. graph white black hispanic other, pie
        title(Ethnic Composition of 26 Housing Projects Combined)
```

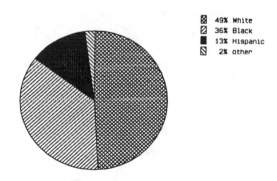

Ethnic Composition of 26 Housing Projects Combined

**Figure 3.21**

Note that all the numerical information in this graph is contained in the legend at upper right: 49% are white, 36% are black, and so on. Pie charts like Figure 3.21 may help to dress up reports, but are useless for analytical purposes—they contain no information that is not more easily read from a table. Using the **by()** option, we could combine multiple pie charts into one image. Such multiple-pie images have better analytical potential, as a tool for quickly scanning for multivariate similarities among cases. Star charts (below) provide a more flexible, if less familiar, tool for this purpose.

Star charts depict multivariate information on a two-dimensional paper or screen. Small figures represent each case, with lines radiating to a distance proportional to the relative value of each variable. For example, *micro.dta* contains data on 51 brands of microcomputers. Variables called *disk*, *clock*, *nop*, *mix*, *fp*, and *ram* measure different aspects of computing speed. Computer brands are stored in the variable *make*:

```
Contains data from c:\stustata\micro.dta
  Obs:      51 (max=  2620)            PC Magazine's benchmark tests
  Vars:      9 (max=    99)
Width:      33 (max=   200)
    1. make         byte    %8.0g    mk    Manufacturer
    2. model        float   %9.0g    md    Model
    3. chip         float   %9.0g          CPU chip type
    4. disk         float   %9.0g          DOS disk access time
    5. clock        float   %9.0g          Clock speed in MHz
    6. nop          float   %9.0g          No operation loop
    7. mix          float   %9.0g          CPU instruction mix
    8. fp           float   %9.0g          Floating point test
    9. ram          float   %9.0g          RAM read/write test
Sorted by:  chip
```

Most of these variables are indexes, so that higher values mean a *slower* computer. A star chart showing six variables for 17 of these computers is produced by (Figure 3.22):

```
. graph disk clock nop mix fp ram if chip==80386 & ram~=., star label(make)
      title(Star Chart of Speed Tests on 21 Personal Computers)
```

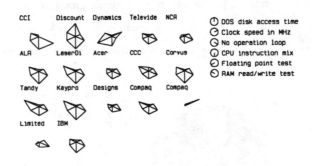

Star Chart of Speed Tests on 21 Personal Computers

**Figure 3.22**

The qualifier **if chip==80386 & ram~=.** instructs Stata to graph cases "only if *chip* equals 80386 [a specific type of computer chip] and if *ram* is not equal to missing."

A long line segment in the *disk* direction represents a high value on this variable; a high value means that disk access is slow. The same holds for the other five variables: the slower the computer, the longer the corresponding line segment. It is hard to read the values of individual variables from a star chart. Instead we look for general patterns. For example, the second Compaq and the Limited stand out as fast machines overall, and the Discount and Laser are generally slow. Several computers have quite similar performance profiles, such as the Laser and Kaypro, or the Televideo and Acer. Note that we are comparing these 17 computers on six variables at once; the chart contains a lot of information. The **by** option does not work with star charts.

## Saving, Printing, and Combining Graphs (Professional Stata Only)

This remainder of this chapter applies only to Professional Stata, not Student Stata. If you just have Student Stata, you may want to skip ahead.

To print a Stata graph, first save it as a disk file by adding **saving(**_drive:filename_**)** to the **graph** command. For example, I originally created Figure 3.6 (page 37) with this command:

```
. graph blue year, connect(1) sort saving(figure6)
```

That saved the graph as a file called _figure6.gph_ (Stata automatically adds a _.gph_ extension) on the current drive and directory. To save _figure6.gph_ on a floppy disk in the A drive:

```
. graph blue year, connect(1) sort saving(a:figure6)
```

or in a directory called _sws_ on the D drive:

```
. graph blue year, connect(1) sort saving(d:\sws\figure6)
```

Once the graph has been saved to disk, DOS, OS/2, and Unix users run a printing program, **gphdot** or **gphpen**. (Mac users may want to skip this section; you just double-click on the graph file's icon to print.) Either program must be told what type of printer we have. The easiest way to do this is to copy the appropriate Stata-supplied driver into a file called `default.dot` (dot matrix or most lasers) or `default.pen` (pen plotters, PostScript, pic files). See the _Stata Reference Manual_ for details about installation. Assuming such installation has already been performed, that we have a dot-matrix or laser printer, and that _figure6.gph_ was saved earlier by **graph** on a disk in drive A, we could print it by three different routes:

1.  If we have enough memory, the easiest way is to run **gphdot** directly from Stata:

    ```
    . gphdot a:figure6
    ```

2.  We might instead use the **shell** command, which also requires memory but gives temporary access to DOS (for file copying, deleting, renaming, or whatever):

    ```
    . shell
    C:\STATA>gphdot a:figure6
    C:\STATA>exit
    .
    ```

    The **shell** command suspends Stata and temporarily returns us to DOS. We then run the **gphdot** program. After typing **exit**, we return to Stata exactly where we left off.

3.  We can use **gphdot** or **gphpen** at any time, not just during a Stata session. For example, we could exit from Stata and then print:

    ```
    . exit
    C:\STATA>gphdot a:figure6
    ```

    **exit**ing from Stata may be necessary for complicated, memory-hungry print jobs.

**gphdot** and **gphpen** support options controlling size and orientation, line thickness, type of printer, and other aspects of the printed images. A few examples:

```
C:\STATA>gphdot a:figure6 /n
```
prints *figure6* without the Stata logo (**/n**)

```
C:\STATA>gphdot a:figure6 /ll /rx110 /ry150
```
prints *figure6* in landscape mode (**/ll**), resizing it horizontally to 110% of normal (**/rx110**) and vertically to 150% of normal (**/ry150**)

```
C:\STATA>gphdot a:figure6 /depsonlq /t28
```
prints *figure6* using an Epson LQ printer (**/depsonlq** overrides whatever device is specified in file default.dot); "pen 1" draws labels and axes two pixels thick and "pen 2" displays the data eight pixels thick (**/t28**)

The *Stata Reference Manual* gives a complete list of options.

Besides printing graphs, **gphpen** also serves to convert Stata graphs into Lotus "pic" format. Some word processors, like WordPerfect, can incorporate (and print) pic files within a document. Other drawing or graphics-enhancement programs can read and edit pic files, and some film recorders can make slides or prints directly from pic files. To produce a pic file from *figure6.gph*, type:

```
. gphpen figure6 /dpic /n
```

This leaves *figure6.gph* unchanged, but creates a new pic-format version called *figure6.pic*. The **/dpic** option causes **gphpen** to use pic.pen (conversion to pic format) as its "printer" file. **/n** suppresses printing the Stata logo.

Aside from printing, there are several other things we can do with saved graphs. Stata's **graph using** command will retrieve and redisplay them. For example:

```
. graph using a:figure6
```

retrieves *figure6.gph* from the A drive, and shows it onscreen.

The **graph using** command can also combine multiple saved graphs into a single image. We can then add titles and save the image as a new graph. For example (Figure 3.23):

```
. graph using figure1 figure7 figure11 figure21, margin(10) title(Professional
    Stata can combine many saved graphs) saving(figure23)
```

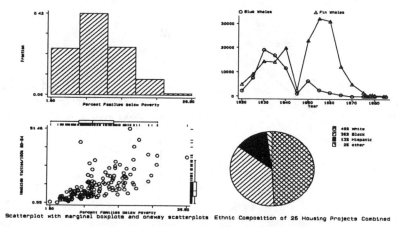

Professional Stata can combine many saved graphs

**Figure 3.23**

In this command *figure1*, *figure7*, ... refer to previously saved Stata graphs in the current drive\directory. Only four graphs are combined in Figure 3.23, but Professional Stata can combine many more. The **margin(10)** option specifies margins between graphs about 10% larger than they otherwise would be. This helps visually separate the individual graphs.

**graph using** arranges individual graphs in a matrix, each graph equally large. The next section points the way to achieving more sophisticated layouts, incorporating text, and editing details of Stata graphs.

## Stata Graphics Editor (Stage)

Stage, the **Stata Graphics Editor**, is a separate program available from Stata's authors at the Computing Resource Center. It allows further refinements of Stata images, such as:

> rearranging, overlaying, or resizing multiple graphs
>
> adding and deleting text or graphical elements
>
> controlling pens and hence color
>
> changing line appearance (dotted, dashed, etc.)
>
> changing plotting symbols (circles, squares, etc.)
>
> changing shading of histograms, bar charts, or pie charts
>
> drawing lines or arrows

Stage-modified graphs like Figure 3.24 can be printed by **gphdot** or **gphpen**, or retrieved and combined by **graph using**, like any other Stata graph.

Users accustomed to the mouse-and-menu interfaces of other drawing programs may at first be put off by Stage's command-line approach, but it is actually simple to learn and use. Unlike a general-purpose graphics editor, Stage understands statistical primitives. For example, Stage initially sees a scatterplot's point cloud as a single entity. Operations like erasing, changing color, or changing plotting symbols can apply to the cloud as a whole. One level deeper, Stage recognizes individual data points, and can edit them one at a time. Stage is also aware of invisible-symbol points and points representing two or more cases with the same location.

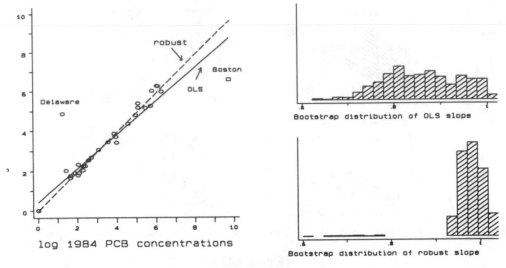

**Figure 3.24**

Anyone preparing Stata-based graphs for publication may find Stage indispensable.  I used Stata and Stage to draw all the original figures for *Regression with Graphics*.

## Also Type help

| | |
|---|---|
| cox | survival analysis graphs |
| diagplots | distributional diagnostic plots |
| egen | generate running averages or medians for smoothing |
| gphdot | dot-matrix or laser printing |
| gphpen | pen plotting, PostScript, or translate to pic format |
| graph | general graphing command |
| greigen | eigenvalue or scree graph (factoring) |
| gr3 | 3-dimensional scatterplot |
| ksm | lowess smoothing |
| ladder | searches ladder of powers for normalizing transformation |
| lv | letter-values display; skewness and elongation diagnostic graphs |
| plot | text-mode scatterplots |
| qc | quality control graphs |
| range | generate numerical range for graphing, derivatives, integrals |
| serrbar | standard-error bar chart |
| stem | stem-and-leaf display |
| survival | survival analysis graphs |
| window | graphics windows (Unix versions only) |

# 4
# Frequency Distributions and Univariate Statistics

Stata's **summarize** command obtains descriptive statistics (mean, standard deviation, etc.) for individual measurement variables. Other measurement-variable procedures include **ci** (confidence intervals for mean), **lv** (letter values), and **means** (geometric and harmonic means). **tabulate** obtains complete frequency distributions, often useful with categorical variables. Options make **tabulate** useful in examining relations between two or more categorical variables (crosstabulations, $\chi^2$ tests, and measures of association) or comparing sample means of one variable across categories of other variables.

## Summarizing a Measurement Variable

Williamstown is a small community in the Green Mountains of Vermont. In 1983, routine state testing detected trace amounts of toxic wastes in the town's water supply. Higher concentrations were found in several private wells and near the town's public schools. Worried citizens formed a Health and Safety Committee (HSC) to press for a solution to this problem. Dataset *toxic.dta* contains information from a 1983 survey of Williamstown residents (Hamilton 1985).

```
. describe

Contains data from c:\stustata\toxic.dta
  Obs:    153 (max=  2620)              Hamilton (1985)
  Vars:     8 (max=    99)
Width:      8 (max=   200)
  1. close          byte    %8.0g   close    schools should close
  2. lived          byte    %8.0g            years lived in Williamstown
  3. educ           byte    %8.0g            highest year school completed
  4. contam         byte    %8.0g   contamlb believe own property/water cont
  5. hsc            byte    %8.0g   kidlbl   Attend 2+ HSC meetings?
  6. female         byte    %8.0g   sexlbl   respondent sex (female)
  7. kids           byte    %8.0g   kidlbl   have kids <19 in Williamstown?
  8. nodad          byte    %8.0g   kidlbl   male non-parent
Sorted by:
```

To find the mean and standard deviation of the variable *lived* (years the respondent had lived in Williamstown) use **summarize**:

```
. summarize lived

Variable |     Obs        Mean   Std. Dev.       Min        Max
---------+-----------------------------------------------------
   lived |     153    19.26797    16.95466         1         81
```

**summarize** also provides the number of observations, minimum, and maximum. For more detailed summary statistics, add the **detail** option:

```
. summarize lived, detail
```

```
                      years lived in Williamstown
-------------------------------------------------------------
        Percentiles      Smallest
  1%         1              1
  5%         2              1
 10%         3              1          Obs                  153
 25%         5              1          Sum of Wgt.          153

 50%        15                         Mean             19.26797
                          Largest      Std. Dev.        16.95466
 75%        29             65
 90%        42             65          Variance         287.4606
 95%        55             68          Skewness         1.208804
 99%        68             81          Kurtosis         4.025642
```

The **summarize, detail** output includes basic statistics plus the following:

Sample percentiles: notably the median (50th percentile), first quartile (25th percentile), and third quartile (75th percentile). These percentiles are only approximations, that work best in large samples.

Four smallest and four largest values, where outliers may show up.

Sum of weights:    Stata understands four types of weights—analytical weights (**aweight**), frequency weights (**fweight**), importance weights (**iweight**), and sampling weights (**pweight**). Different procedures allow, and make sense with, different kinds of weights. **summarize**, for example, permits only **aweight** or **fweight**. Type **help weight** for explanations.

Variance:    standard deviation squared (more properly, standard deviation equals the square root of variance).

Skewness:    the direction and degree of asymmetry. A perfectly symmetrical distribution has skewness = 0. Positive skew (heavier right tail) produces skewness > 0; negative skew (heavier left tail) produces skewness < 0.

Kurtosis:    tail heaviness. A normal (Gaussian) distribution has kurtosis = 3. If kurtosis > 3, the distribution is heavy-tailed (or sharply peaked) relative to a normal distribution. Kurtosis < 3 indicates lighter-than-normal tails.

**ci** calculates confidence intervals for the mean. Here is a 99% confidence interval for the mean length of residence (*lived*) for the Williamstown survey respondents:

```
. ci lived, level(99)

Variable |     Obs        Mean     Std. Err.      [99% Conf. Interval]
---------+-----------------------------------------------------------
   lived |     153     19.26797    1.370703       15.69241    22.84354
```

The **level()** option specifies the desired degree of confidence. If we simply typed **ci lived**, we would get a 95% confidence interval. Another option, **poisson**, would produce an exact confidence interval for a Poisson-distributed count variable.

Many statistical procedures assume normally distributed variables. To check on the reasonableness of this assumption, we should look closely at the data. Stem-and-leaf displays provide a quick visual check, most useful with small to medium-sized samples. Initial digits form the "stems" of a stem-and-leaf display; following digits for each case constitute its "leaves." For example, still using years resident in Williamstown (*lived*):

```
. stem lived

Stem and leaf plot for lived

  0 | 11111112222233333333444444445555555555556666666677889999
  1 | 00000011222233333345555555567788899
  2 | 000000111112224444456778899
  3 | 000001245555666789
  4 | 001259
  5 | 00134556
  6 | 5558
  7 |
  8 | 1
```

Letter-value displays (**lv**) use order statistics to dissect a distribution.  Creative statistician John Tukey (1977) gets credit for the invention and unique terminology of both stem-and-leaf and letter-values displays.

```
. lv lived
```

| # | 153 | | years lived in Williamstown | | | | |
|---|---|---|---|---|---|---|---|
| | | | --------------------------------- | | | spread | pseudosigma |
| M | 77 | | | 15 | | | |
| F | 39 | | 5 | 17 | 29 | 24 | 17.9731 |
| E | 20 | | 3 | 21 | 39 | 36 | 15.86391 |
| D | 10.5 | | 2 | 27 | 52 | 50 | 16.62351 |
| C | 5.5 | | 1 | 30.75 | 60.5 | 59.5 | 16.26523 |
| B | 3 | | 1 | 33 | 65 | 64 | 15.15955 |
| A | 2 | | 1 | 34.5 | 68 | 67 | 14.59762 |
| Z | 1.5 | | 1 | 37.75 | 74.5 | 73.5 | 15.14113 |
| | 1 | | 1 | 41 | 81 | 80 | 15.32737 |
| | | | | | | # below | # above |
| inner fence | | | -31 | | 65 | 0 | 5 |
| outer fence | | | -67 | | 101 | 0 | 0 |

$M$ denotes the median, and $F$ the "fourths" (quartiles, using a different approximation than **summarize, detail** does).  $E, D, C, \ldots$ denote cutoff points such that roughly $1/8$, $1/16$, $1/32, \ldots$ of the distribution remains outside in the tails.  The second column of numbers gives the depth (distance from nearest extreme) for each letter value.  Within the center box, the middle column gives "midsummaries," simply averages of the two letter values.  If midsummaries drift away from the median (as they do for *lived*) the distribution becomes progressively more skewed as we move farther out into the tails.  The "spreads" are differences between pairs of letter values; the spread between $F$'s equals approximate interquartile range, for instance.  Finally, the right-hand column of "pseudosigmas" estimates what the standard deviation should be if these letter values described a Gaussian population.  The $F$ pseudosigma, sometimes called a "pseudo standard deviation" (*PSD*), provides a simple check for normality in symmetrical distributions:

1.  Comparing mean with median diagnoses overall skew:

    mean > median      positive skew
    mean ≈ median      symmetry
    mean < median      negative skew

2.  If mean and median indicate symmetry, comparing standard deviation with *PSD* evaluates tail normality:

    standard deviation > *PSD*      heavier-than-normal tails
    standard deviation ≈ *PSD*      normal tails
    standard deviation < *PSD*      lighter-than-normal tails

Let $F_1$ and $F_3$ denote 1st and 3rd fourths (approximate 25th and 75th percentiles).  Then the interquartile range, *IQR*, equals $F_3 - F_1$, and $PSD = IQR/1.349$.

lv  also identifies mild and severe outliers.    We call $x$ a "mild outlier" if:

$$F_1-3IQR \leq x < F_1-1.5IQR \quad \text{or} \quad F_3+1.5IQR < x \leq F_3+3IQR$$

$x$ is a "severe outlier" if:

$$x < F_1-3IQR \quad \text{or} \quad x > F_3+3IQR$$

lv  gives these cutoffs, called "inner fences" and "outer fences," and the number of outliers beyond them.  Severe outliers, cases beyond the outer fences, occur sparsely (about two per million) in normal populations.  Monte Carlo simulations suggest that the presence of any severe outliers in samples of $n = 15$ to about 20,000 should be sufficient evidence to reject a normality hypothesis at $\alpha = .05$ (Hamilton 1992b).  Furthermore, severe outliers represent the kind of nonnormality most damaging to many statistical techniques.

summarize, stem, and  lv  all show that *lived* has a positively skewed, clearly nonnormal sample distribution.  For a formal test of the normal-population hypothesis, we could employ sktest (skewness and kurtosis test of normality):

. sktest *lived*

```
                   Skewness/Kurtosis tests for Normality
                                            ------- joint -------
   Variable |  Pr(Skewness)   Pr(Kurtosis)  adj chi-sq(2)  Pr(chi-sq)
 ----------+---------------------------------------------------------
     lived |      0.000         0.028          24.79        0.0000
```

sktest here rejects normality: *lived* appears significantly nonnormal in skewness ($P = .000$), kurtosis ($P = .028$), and both considered jointly ($P = .0000$).  Stata rounds off displayed probabilities to three or four decimals; "$P = 0.0000$" really means $P < .00005$.

## Frequencies and Crosstabulation

The methods described above apply to measurement variables.  Categorical variables require other approaches, such as tabulation.  To find the percentage of Williamstown respondents attending two or more Health and Safety Committee meetings, tabulate the categorical variable *hsc*:

. tabulate *hsc*

```
 Attend 2+|
      HSC|
 meetings?|     Freq.      Percent        Cum.
 ---------+------------------------------------
      no |       106        69.28        69.28
     yes |        47        30.72       100.00
 ---------+------------------------------------
   Total |       153       100.00
```

tabulate can also handle measurement variables, if they do not have too many values.

tabulate followed by two variable names performs crosstabulation.  For example, here is a crosstabulation of *hsc* by *kids* (whether respondent has children under 19 living in town):

. tabulate *hsc kids*

```
 Attend 2+| have kids <19 in Williamstown?
      HSC|
 meetings?|      no         yes  |     Total
 ---------+---------------------+----------
      no |      52          54  |       106
     yes |      11          36  |        47
 ---------+---------------------+----------
   Total|      63          90  |       153
```

The first-named variable forms the rows, and the second forms columns in the resulting crosstabulation. We see that only 11 of these 153 people were non-parents attending the HSC meetings.

Several options enhance crosstabulation:

| | |
|---|---|
| column | column percentages |
| row | row percentages |
| cell | total percentages |
| nofreq | do not print cell frequencies |
| chi2 | $\chi^2$ test of independence |
| lrchi2 | likelihood-ratio $\chi^2$ |
| exact | Fisher's exact test |
| V | Cramer's $V$ |
| gamma | Goodman & Kruskall's $\gamma$ |
| taub | Kendall's $\tau_b$ |
| all | all the above |

For example, to get column percentages (because the column variable, *kids*, is the independent variable in this analysis) and a $\chi^2$ test:

```
. tabulate hsc kids, column chi2

Attend 2+| have kids <19 in Williamstown?
    HSC|
meetings?|        no        yes |     Total
----------+----------------------+----------
      no |        52         54 |       106
         |     82.54      60.00 |     69.28
----------+----------------------+----------
     yes |        11         36 |        47
         |     17.46      40.00 |     30.72
----------+----------------------+----------
   Total|        63         90 |       153
         |    100.00     100.00 |    100.00

        Pearson chi2(1) =    8.8464    Pr = 0.003
```

We see a statistically significant ($P = .003$) relationship. Forty percent of the respondents with children attended HSC meetings, compared with only 17.46% of the respondents without children.

With the **by** prefix, **tabulate** produces multiway contingency tables. Here is a 3-way crosstabulation of *hsc* by *kids* by *contam* (respondent's own property or water contaminated?):

```
. sort contam
. by contam:  tabulate hsc kids, column

-> contam=      no
 Attend 2+| have kids <19 in Williamstown?
    HSC|
meetings?|        no        yes |     Total
----------+----------------------+----------
      no |        42         44 |        86
         |     91.30      68.75 |     78.18
----------+----------------------+----------
     yes |         4         20 |        24
         |      8.70      31.25 |     21.82
----------+----------------------+----------
   Total|        46         64 |       110
         |    100.00     100.00 |    100.00
```

```
-> contam=      yes
  Attend 2+| have kids <19 in Williamstown?
      HSC|
  meetings?|        no        yes  |     Total
-----------+----------------------+----------
        no |        10         10  |        20
           |     58.82      38.46  |     46.51
-----------+----------------------+----------
       yes |         7         16  |        23
           |     41.18      61.54  |     53.49
-----------+----------------------+----------
     Total |        17         26  |        43
           |    100.00     100.00  |    100.00
```

The "parenthood effect" shows up among both contaminated and uncontaminated respondents.

This approach extends to tabulations of any complexity. For example, to get a 4-way crosstabulation of *female* by *contam* by *hsc* by *kids* (results not shown):

```
. sort female contam
. by female contam:  tabulate hsc kids, column
```

For a 5-way table of *close* by *female* by *contam* by *hsc* by *kids* (results not shown):

```
. sort close female contam
. by close female contam:  tabulate hsc kids, column
```

and so forth. Formally modeling multiway contingency tables requires multivariate techniques such as loglinear modeling or logistic regression (type **help loglin**, or **help logit**).

## Tables of Means

**tabulate** will also produce tables of means and standard deviations within categories of the tabulated variable. For example, to find mean years resident (*lived*) for respondents who did and did not attend HSC meetings:

```
. tabulate hsc, summ(lived)

  Attend 2+| Summary of years lived in Williamstown
      HSC|
  meetings?|      Mean    Std. Dev.       Freq.
-----------+------------------------------------
        no | 21.509434   17.743809         106
       yes | 14.212766   13.911109          47
-----------+------------------------------------
     Total | 19.267974   16.954663         153
```

HSC attenders appear to be relative newcomers, averaging 14.2 years in Williamstown.

Means can also be displayed within a two-variable table:

```
. tabulate hsc kids, summ(lived) means

              Means of years lived in Williamstown

  Attend 2+| have kids <19 in Williamstown?
      HSC|
  meetings?|        no        yes       Total
-----------+----------------------+----------
        no | 28.307692  14.962963  | 21.509434
       yes | 23.363636  11.416667  | 14.212766
-----------+----------------------+----------
     Total | 27.444444  13.544444  | 19.267974
```

Both parents and nonparents among the HSC attenders tend to have lived fewer years in Williamstown, so the newcomer/oldtimer division noticed in the previous table is not a spurious result of parenthood.

The **means** option calls for a tabulation containing only means. Otherwise we get a bulkier table including means, standard deviations, and frequencies in each cell. Chapter 5 describes a variety of procedures for detecting significant differences between means.

## Multiple Oneway or Twoway Tables

With surveys and other large datasets, we often want frequency distributions for many different variables. Instead of asking for each table separately, for example by typing **tabulate** *hsc*, then **tabulate** *female*, and then **tabulate** *kids*, we could simply use **tab1**:

. tab1 *hsc female kids*

or, to list frequencies of every variable in this dataset:

. tab1 *close-nodad*

**tab1** obtains oneway frequency tables for up to 30 variables at a time.

**tab2** provides similar capabilities with twoway tables (crosstabulations). For example,

. tab2 *hsc female kids*

obtains crosstabulations of every possible combination of the listed variables. Finally, **tab3** (supplied with Student Stata) offers crosstabulations of the first-named variables by all other variables specified. For example,

. tab3 *hsc female kids*

would crosstabulate *female* by *hsc*, then *kids* by *hsc*. The usual **tabulate** options work as expected with **tab1**, **tab2**, and **tab3**.

## Also type help

| | |
|---|---|
| **alpha** | Cronbach's $\alpha$ reliability |
| **boxcox** | maximum-likelihood Box-Cox normalizing transformations |
| **ci** | means with confidence intervals |
| **egen** | generate means, medians, running averages, differences, etc. as variables |
| **kappa** | interrater reliability |
| **lv** | Tukey letter-values display |
| **means** | arithmetic, geometric, and harmonic means |
| **sktest** | skewness and kurtosis test for normality |
| **stem** | stem-and-leaf display |
| **summarize** | mean, standard deviation, and other summary statistics |
| **swilk** | Shapiro-Wilk and Shapiro-Francia tests for normality |
| **tabsum** | tables of means |
| **tabulate** | frequency tables, crosstabulations, tables of means |
| **tab1** | many oneway frequency tables, one for each specified variable |
| **tab2** | many twoway crosstabulations, all combinations of specified variables |

| tab3 | many twoway crosstabulations, first-named variable by other specified variables (supplied with Student Stata) |
| weight | case weights:  analytical, frequency, importance, or sampling |

# 5
# *t* Tests, ANOVA, and Nonparametric Comparisons

Stata offers a variety of methods for comparing group centers. These methods fall into two broad types:

1. "parametric" methods (such as **ttest**, **oneway**, or **anova**) that test for differences between means; or

2. "nonparametric" methods (such as **signrank**, **ranksum**, or **kwallis**) that test for differences between medians.

Parametric tests tend to be more powerful than nonparametric tests if their distributional assumptions are met. Furthermore, they form a unified system extending from univariate to multivariate analysis. (Many parametric mean-comparison methods are special cases of ANOVA or regression, and could be performed using Stata's versatile **anova** or **regress** commands.) On the other hand, nonparametric methods require fewer assumptions, and they are generally robust against problems like outliers and nonconstant variance. A careful analyst might try both approaches and feel encouraged if they point to similar conclusions. Investigate further if parametric and nonparametric results disagree.

## Paired-Difference Tests

Paired-difference tests compare two variables, across the same set of cases. For example, the data in *writing.dta* were collected to evaluate a college writing course that employed microcomputers for word processing (Nash and Schwartz 1987). Measures such as number of sentences completed in timed writing were collected both before and after students took the course. The researchers were interested in whether the post-course measures showed improvement.

```
. describe

Contains data from c:\stustata\writing.dta
  Obs:     24 (max=  2620)              Nash and Schwartz (1987)
  Vars:     9 (max=    99)
 Width:     9 (max=   200)
   1. student     byte    %8.0g    slbl    Student ID
   2. preS        byte    %8.0g            Number of Sentences (pretest)
   3. preP        byte    %8.0g            Number of Paragraphs (pretest)
   4. preC        byte    %8.0g            Coherence Scale 0-2 (pretest)
   5. preE        byte    %8.0g            Evidence Scale 0-6 (pretest)
   6. postS       byte    %8.0g            Number of Sentences (postest)
   7. postP       byte    %8.0g            Number of Paragraphs (postest)
   8. postC       byte    %8.0g            Coherence Scale 0-2 (postest)
   9. postE       byte    %8.0g            Evidence Scale 0-6 (postest)
Sorted by:
```

To obtain a paired-difference *t* test for a change in mean number of sentences completed:

```
. ttest preS = postS
```

```
Variable |      Obs        Mean    Std. Dev.
---------+-------------------------------------
   preS |       24     10.79167    4.606037
  postS |       24       26.375    8.297787
---------+-------------------------------------
   diff. |       24    -15.58333    6.775382

        Ho:  diff = 0   (paired data)
                 t = -11.27 with 23 d.f.
        Pr > |t| = 0.0000
```

The notation `Prob > |t|` means "the probability of a greater absolute value of *t*." Since this probability is low (*P* = .0001), the post-course mean (26.4) differs significantly from the pre-course mean (10.8). **ttest** automatically performs two-tailed tests; for a one-tailed test, divide the output *P*-value in half (*P* = .0001/2 = .00005).

Notice that the post-course standard deviation (8.3) is almost twice the pre-course standard deviation (4.6). This casts doubt on the *t* test's constant-variance assumption. Perhaps we should instead use a nonparametric procedure (Wilcoxon signed-rank test) that does not require this assumption:

. **signrank** *preS = postS*

```
Test:  Equality of distributions (Wilcoxon Signed-Ranks)

Result of preS - (postS)
  Sum of Positive Ranks = 0
  Sum of Negative Ranks = 300

  z-statistic   -4.29
  Prob > |z|    0.0001
```

**signrank** finds significantly different medians. Whereas **ttest**'s paired-difference test assumes normal and equal-variance population distributions, **signrank** assumes only that they are symmetrical and continuous. Since both tests confirm a difference between the centers of pre- and post-course distributions, we can state this finding more confidently.


## Two-Sample Tests

The remainder of this chapter uses data from a survey of college undergraduates. Student Stata provides these data in two files: *student1.dta* has the complete dataset (*n* = 243), but it is too large to fit into Student Stata. File *student2.dta* contains a random subsample (*n* = 150) that can fit into Student Stata. For illustration, I use the complete version.

. **describe**

```
Contains data from c:\stustata\student1.dta
  Obs:    243 (max=  1473)               Sociology students fall 1987
  Vars:    20 (max=    99)
  Width:   50 (max=   200)
    1. id1          int     %8.0g              case number
    2. year         int     %8.0g    v1        year in college
    3. age          int     %8.0g              Age at last birthday
    4. major        int     %8.0g              Student major
    5. relig        int     %8.0g    v4        Religious preference
    6. gpa          float   %9.0g              Grade Point Average
    7. grades       int     %8.0g    grades    Guessed Grades this Semester
    8. greek        int     %8.0g    greek     Member Fraternity or Sorority?
    9. socsci       int     %8.0g              # Social Science Courses Taken?
   10. live         int     %8.0g    v10       Where do you live?
   11. miles        int     %8.0g              How many miles from campus?
   12. study        int     %8.0g              Avg. hours/week studying
```

```
13. athlete     int    %8.0g    yes      Are you a varsity athlete?
14. employed    int    %8.0g    yes      Are you employed
15. allnight    int    %8.0g    allnight How often study all night?
16. ditch       int    %8.0g    times    How many class/month ditched?
17. drink       float  %9.0g             33-point drinking scale
18. hsdrink     float  %9.0g             High School Drinking Scale
19. aggress     float  %9.0g             Aggressive behavior scale
20. sex         float  %9.0g    s        Sex (male)
Sorted by:  id1
```

About 19% of these students belong to a fraternity or sorority:

```
. tabulate greek

    Member|
 Fraternity|
or Sorority?|       Freq.       Percent       Cum.
-----------+-----------------------------------------
     Greek |          47         19.34        19.34
  NonGreek |         196         80.66       100.00
-----------+-----------------------------------------
     Total |         243        100.00
```

The dichotomy *greek* has values of 1 ("Greek") or 2 ("NonGreek"), as revealed if we **tabulate** with the **nolabel** option:

```
. tabulate greek, nolabel

    Member|
 Fraternity|
or Sorority?|       Freq.       Percent       Cum.
-----------+-----------------------------------------
         1 |          47         19.34        19.34
         2 |         196         80.66       100.00
-----------+-----------------------------------------
     Total |         243        100.00
```

Another variable, *drink*, is a 33-point drinking scale. Are Greeks and non-Greeks significantly different in their drinking behavior? We might answer this question using a two-sample *t* test for a difference of means. **ttest** will perform such a test. Its general syntax, for a two-sample test, is **ttest *measurement*, by(*categorical*)**. For example:

```
. ttest drink, by(greek)

Variable |      Obs        Mean    Std. Dev.
---------+-----------------------------------
       1 |       47     24.7234    4.884323
       2 |      196     17.7602    6.405018
---------+-----------------------------------
combined |      243      19.107    6.722117

        Ho:  mean(x) = mean(y)    (assuming equal variances)
                  t = 6.98 with 241 d.f.
          Pr > |t| = 0.0000
```

The output notes that these results rest on an equal-variance assumption. But the fraternity/sorority members' standard deviation is somewhat lower—they are more alike than non-Greeks, in terms of their drinking behavior. **ttest** can also perform a two-sample test without the equal-variance assumption if we add the option **unequal**:

```
. ttest drink, by(greek) unequal
```

```
Variable |      Obs        Mean     Std. Dev.
---------+-----------------------------------
      1 |       47     24.7234     4.884323
      2 |      196     17.7602     6.405018
---------+-----------------------------------
combined |      243     19.107
```

$$Ho: \quad mean(x) = mean(y) \quad (assuming \ unequal \ variances)$$
$$t = 8.22 \ with \ 90 \ d.f.$$
$$Pr > |t| = 0.0000$$

This adjustment does not alter our finding of a significant difference. Alternatively, we might apply a nonparametric procedure, the Wilcoxon rank sum test (also called the Mann-Whitney $U$ test). ranksum's syntax resembles ttest's:

. ranksum *drink*, by(*greek*)

```
Test: Equality of medians (Two-Sample Wilcoxon Rank-Sum)

Sum of Ranks: 8535 (greek == 1)
Expected Sum: 5734

z-statistic  6.47
Prob > |z|   0.0001
```

The rank sum test rejects a null hypothesis of equal medians ($P = .0001$). This test assumes only that the population distributions have similar shapes.

Oneway analysis of variance (ANOVA) provides another way to test for differences between means. For example:

. oneway *drink greek*, tabulate

```
     Member|   Summary of 33-point drinking scale
  Fraternity|
or Sorority?|    Mean     Std. Dev.        Freq.
-----------+-----------------------------------------
     Greek |  24.723404   4.8843233          47
  NonGreek |  17.760204   6.4050179         196
-----------+-----------------------------------------
     Total |  19.106996   6.7221166         243
```

```
                     Analysis of Variance
   Source           SS          df      MS            F      Prob > F
---------------------------------------------------------------------
Between groups   1838.08426      1   1838.08426    48.69     0.0000
Within groups    9097.13385    241   37.7474433
---------------------------------------------------------------------
   Total        10935.2181     242   45.1868517
```

Bartlett's test for equal variances:  chi2(1) =   4.8378  Prob>chi2 = 0.028

The tabulate option produces a table of means and standard deviations, printed above usual oneway ANOVA results. This oneway ANOVA with a dichotomous $X$ variable (oneway drink greek) is mathematically equivalent to a two-sample $t$ test (ttest drink, by(greek)). oneway offers more options and executes faster, but lacks ttest's unequal option for abandoning the equal-variances assumption.

oneway does formally test the equal-variances assumption, using Bartlett's $\chi^2$. A low Bartlett's test $P$-value implies that ANOVA's equal-variance assumption is implausible, so the ANOVA ($F$ test) results may be untrustworthy. In the oneway drink greek example, Bartlett's $P = .028$, casting doubt on the ANOVA finding of a significant ("P = .0000," actually meaning $P < .00005$) difference between means.

If unequal variances or nonnormal distributions undermine basic ANOVA or *t* test results, we have two simple alternatives:

1. **ttest, unequal** provides a two-sample *t* test without assuming equal variances.

2. **ranksum** provides a two-sample nonparametric test, not assuming either equal variances or normality.

Two less-simple alternatives are also worth mentioning:

3. Power transformations (Chapter 2) may succeed in stabilizing variance and/or normalizing distributions.

4. Robust estimators (Chapter 10) do not require normality.

Alternatives 3 and 4 become increasingly attractive for more complex analyses.

## Oneway Analysis of Variance

The previous section introduced oneway ANOVA as a method for comparing two means, but in fact it can compare any number of means. Using the same undergraduate survey data (*student1.dta*), we might test whether mean drinking behavior changes with year in college:

```
. oneway drink year, tabulate scheffe
```

| year in college | Summary of 33-point drinking scale Mean | Std. Dev. | Freq. |
|---|---|---|---|
| Freshman | 18.975 | 6.9226033 | 40 |
| Sophomor | 21.169231 | 6.5444853 | 65 |
| Junior | 19.453333 | 6.2866081 | 75 |
| Senior | 16.650794 | 6.6409257 | 63 |
| Total | 19.106996 | 6.7221166 | 243 |

Analysis of Variance

| Source | SS | df | MS | F | Prob > F |
|---|---|---|---|---|---|
| Between groups | 666.200518 | 3 | 222.066839 | 5.17 | 0.0018 |
| Within groups | 10269.0176 | 239 | 42.9666008 | | |
| Total | 10935.2181 | 242 | 45.1868517 | | |

Bartlett's test for equal variances:  chi2(3) =  0.5103  Prob>chi2 = 0.917

Comparison of 33-point drinking scale by year in college
(Scheffe)

| Row Mean- Col Mean | Freshman | Sophomor | Junior |
|---|---|---|---|
| Sophomor | 2.19423 0.429 | | |
| Junior | .478333 0.987 | -1.7159 0.498 | |
| Senior | -2.32421 0.382 | -4.51844 0.002 | -2.80254 0.103 |

The **tabulate** option asks for a table of means, in which we see that Sophomores drink the most, followed by Juniors, Freshmen, and Seniors. The ANOVA *F* test ($F = 5.17$, $P = .0018$) leads to rejecting the null hypothesis that the four groups' population means are equal. Bartlett's test ($\chi^2 = .5103$, $P = .917$) gives us no reason to doubt the equal-variances assumption upon which ANOVA rests.

The **scheffe** option (Scheffé multiple-comparison test) produces a table showing differences between each pair of means. For example, the freshman mean is 18.975 and the sophomore mean is 21.16923, so the sophomore-freshman difference is 21.16923 − 18.975 = 2.19423, not statistically distinguishable from zero ($P = .429$). Of the six contrasts in this table, only the senior-sophomore difference, 16.6508 − 21.1692 = −4.5184, is significant ($P = .002$). Thus our overall finding that these four groups' means are not the same arises mainly from the contrast between seniors (the lightest drinkers) and sophomores (the heaviest).

**oneway** offers three multiple-comparison options: **scheffe**, **bonferroni**, or **sidak**. The Scheffé test remains valid under a wider variety of conditions, though it is sometimes less sensitive.

The Kruskal-Wallis test (**kwallis**), a $K$-sample generalization of the two-sample rank sum test (**ranksum**), provides a nonparametric alternative to oneway ANOVA. It tests the null hypothesis of identical population medians, rather than means:

```
. kwallis drink, by(year)

Test: Equality of populations (Kruskal-Wallis Test)

     year     _Obs    _RankSum
  Freshman      40     4914.00
  Sophomor      65     9341.50
    Junior      75     9300.50
    Senior      63     6090.00

  chi-square =     14.453 with 3 d.f.
  probability =     0.0023
```

Here, the **kwallis** results ($P = .0023$) agree with our **oneway** findings ($P = .0018$). Kruskal-Wallis is generally safer than ANOVA if we have reason to doubt ANOVA's equal-variances or normality assumptions, or suspect problems with outliers. Like **ranksum**, **kwallis** makes the weaker assumption of similar-shaped distributions within each group.

## Error-Bar Charts

Experienced researchers learn to consult graphs of variable distributions at an early stage in data analysis. This helps to screen out potential problems and suggest directions for further analysis. Parallel boxplots provide an ideal graphical companion to the comparison tests described in this chapter. For example, Figure 5.1 shows boxplots corresponding to the previous section's analysis of drinking behavior by year in college:

```
. sort year
. graph drink, box by(year) yline(19) ylabel(5,10,15,20,25,30,35)
      l1(Drinking Scale)
```

A horizontal line in Figure 5.1 marks the overall median (19). The four distributions have similar spreads and appear reasonably symmetrical, with no outliers. Figure 5.1 shows no reason to doubt the equal-variance or normality assumptions underlying the earlier ANOVA.

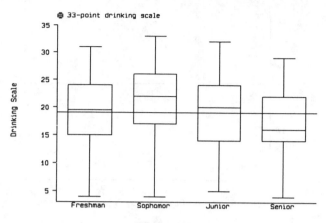

**Figure 5.1**

Research reports often visualize ANOVA results using error-bar charts. Typically these charts show each group's mean, plus and minus one standard error. To construct an error-bar chart, we clear memory and type in a small new dataset consisting of the group means:

```
. clear
. input year mean n
          year         mean          n
  1. 1 18.975 40
  2. 2 21.169 65
  3. 3 19.453 75
  4. 4 16.651 63
  5. end
```

*year* indicates year in college (1=freshman, 2=sophomore, etc.). Drinking-scale means (*mean*) and group size (*n*) are copied from the `oneway, tabulate` output (page 65). Next we estimate the means' standard errors. Divide the square root of the within-groups mean square (42.9666008) by the square root of group size (*n*):

```
. generate SE = sqrt(42.9666008)/sqrt(n)
```

`serrbar` can now draw an error-bar chart showing mean ±*SE* (standard error). A horizontal line shows the grand mean, 19.108 (Figure 5.2):

```
. serrbar mean SE year, yline(19.108) title(Error-Bar Chart of Drinking by
     Year in College) ylabel(15,16,17,18,19,20,21,22) xlabel(1,2,3,4)
```

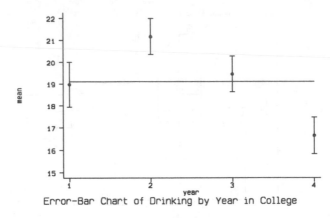

Error-Bar Chart of Drinking by Year in College

**Figure 5.2**

Alternative forms of error-bar charts include:

1.  bars displaying a 95% confidence interval (± about two standard errors, depending on degrees of freedom);

2.  bars extending ± one standard deviation, rather than ± one standard error;

3.  bars showing the entire range of the data within each group, or some other asymmetrical interval; or

4.  line segments connecting group means, to show upwards or downwards trends.

The first two variations involve minor changes in the **serrbar** command line (type **help serrbar** for details) or the underlying dataset of means. The third and fourth variations require more direct control of the graphing, through Stata's **graph** command. We need to explicitly define the high and low points of each interval. For example, to draw a graph like Figure 5.2 but with lines connecting the group means (Figure 5.3):

```
. generate lo = mean-SE
. generate hi = mean+SE
. graph mean hi lo year, connect(lII) symbol(Oii) yline(19.108)
      title(Error-Bar Chart of Drinking by Year in College)
      ylabel(15,16,17,18,19,20,21,22) xlabel(1,2,3,4)
```

This command tells Stata to graph three *Y* variables (*mean*, *hi*, *lo*) against *year*. Connect the first-named variable with line segments (**l**) and the next two with capped vertical bars (**II**). The first-named variable plots with large circles (**O**), and the next two are invisible (**ii**). More complicated error-bar graphs for *N*-way analysis of variance (next section) can also be produced in this fashion, applying **graph**'s versatile **connect()** and **symbol()** options to small datasets of means and standard errors.

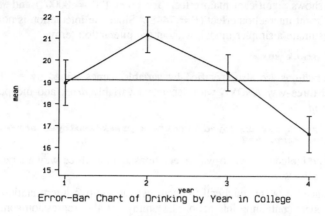

Error-Bar Chart of Drinking by Year in College

**Figure 5.3**

## *N*-Way Analysis of Variance

*N*-way analysis of variance refers to ANOVA with two or more categorical *X* variables. For example, we might examine how drinking behavior varies with fraternity/sorority membership and gender. **tabulate** can produce a table of means:

```
. tabulate greek sex, summarize(drink) means
```

```
                    Means of 33-point drinking scale

    Member| Sex (male)
Fraternity|
       or|
 Sorority?|  Female      Male      Total
----------+--------------------------+----------
    Greek | 22.444444  26.137931 | 24.723404
 NonGreek | 16.517241   19.5625  | 17.760204
----------+--------------------------+----------
    Total | 17.313433  21.311927 | 19.106996
```

It appears that males drink more than females, and fraternity/sorority members drink more than non-members. The male-female difference is slightly greater among non-members. **anova** tests for significant differences among these means attributable to Greek membership, gender, or the interaction of membership and gender (*greek\*sex*):

```
. anova drink greek sex greek*sex
```

```
                      Number of obs =     243    R-square     =  0.2221
                      Root MSE      = 5.96592    Adj R-square =  0.2123

           Source |  Partial SS    df       MS           F     Prob > F
        ----------+----------------------------------------------------
            Model |  2428.67237     3   809.557456       22.75   0.0000
                  |
            greek |   1406.2366     1    1406.2366       39.51   0.0000
              sex |  408.520097     1   408.520097       11.48   0.0008
        greek*sex |  3.78016612     1   3.78016612        0.11   0.7448
                  |
         Residual |  8506.54574   239   35.5922416
        ----------+----------------------------------------------------
            Total |  10935.2181   242   45.1868517
```

The output shows significant main effects for *greek* ("P = .0000") and *sex* (*P* = .0008). There is no significant interaction effect (*P* = .7448). Since the interaction is nonsignificant, we might prefer to estimate a simpler model, without the interaction term:

```
. anova drink greek sex
```

Higher-order interactions are also specified by variable names joined by *. For example, to obtain a factorial three-way ANOVA with dependent variable *drink* and independent variables *greek*, *sex*, and *employed*:

```
. anova drink greek sex employed greek*sex greek*employed sex*employed
     greek*sex*employed
```

This analysis would include three two-way interactions and one three-way interaction, in addition to the three main effects.

With unbalanced data (unequal cell frequencies), substantive interpretation of main effects is awkward in models containing interactions. Dummy variable regression sometimes provides a better way to analyze such data.


## Analysis of Covariance

Unless told otherwise, **anova** assumes that all *X* variables are categorical. To perform an analysis of covariance (ANCOVA), with a mixture of categorical and measurement independent variables, list the measurement variables within a **continuous(*varlist*)** option. For example, does the measurement variable college-grade-point-average (*gpa*) predict drinking behavior?

```
. anova drink greek sex gpa, continuous(gpa)
```

```
               Number of obs =      218     R-square     =  0.2970
               Root MSE      = 5.68939     Adj R-square =  0.2872

     Source |  Partial SS    df       MS              F     Prob > F
-----------+----------------------------------------------------------
      Model |  2927.03087     3   975.676958          30.14    0.0000
            |
      greek |  1489.31999     1   1489.31999          46.01    0.0000
        sex |  405.137843     1   405.137843          12.52    0.0005
        gpa |    407.0089     1     407.0089          12.57    0.0005
            |
   Residual |  6926.99206   214   32.3691218
-----------+----------------------------------------------------------
      Total |  9854.02294   217   45.4102439
```

The answer seems to be yes.

ANOVA or ANCOVA can be viewed as special cases of regression. To make the underlying regression model explicit, add the **regress** option to any **anova** command. For example:

```
. anova drink sex year sex*year gpa, continuous(gpa) regress
```

```
 Source |      SS        df       MS            Number of obs =     218
--------+---------------------------            F(  8,   209) =    6.57
  Model |  1981.3043      8  247.663037         Prob > F      =  0.0000
Residual| 7872.71864    209  37.6685102         R-square      =  0.2011
--------+---------------------------            Adj R-square  =  0.1705
  Total |  9854.02294   217  45.4102439         Root MSE      =  6.1375
```

```
------------------------------------------------------------------------------
   drink       Coef.    Std. Err.      t     P>|t|    [95% Conf. Interval]
------------------------------------------------------------------------------
   _cons     26.97099   2.915372     9.251   0.000    21.22369     32.7183
sex
      1    -3.840237    1.590542    -2.414   0.017    -6.975799   -.7046752
      2    (dropped)
year
      1    -2.154446    2.494539    -0.864   0.389    -7.072129    2.763237
      2     4.083382    1.661489     2.458   0.015     .8079567    7.358807
      3     1.308525    1.608042     0.814   0.417    -1.861536    4.478586
      4    (dropped)
gpa        -2.750319     .945631    -2.908   0.004    -4.614516   -.8861213
sex*year
    1 1     3.346156    3.312956     1.010   0.314    -3.184936    9.877249
    1 2    -1.393867    2.203533    -0.633   0.528    -5.737868    2.950133
    1 3     1.542029    2.137601     0.721   0.471    -2.671994    5.756051
    1 4    (dropped)
    2 1    (dropped)
    2 2    (dropped)
    2 3    (dropped)
    2 4    (dropped)
------------------------------------------------------------------------------
```

With the **regress** option, **anova** organizes its output as a regression table (see Chapters 6-7). The top part gives the same overall $F$ test and $R^2$ as a standard ANOVA table. The bottom part describes the following regression:

> We construct a separate dummy variable {0,1} representing each category of each $X$ variable, except for the highest categories which are dropped. Interaction terms (if specified) are constructed from the products of every possible combination of these dummy variables. Regress $Y$ on all these dummy variables and interactions, and also on any continuous variables in the analysis.

The ANOVA above therefore corresponds to a regression of *drink* on eight $X$ variables:

1.  a dummy coded 1=female, 0 otherwise;
    (highest category of *sex*, male, gets dropped)
2.  a dummy coded 1=freshman, 0 otherwise;
3.  a dummy coded 1=sophomore, 0 otherwise;
4.  a dummy coded 1=junior, 0 otherwise;
    (highest category of *year*, senior, gets dropped)
5.  the continuous variable *gpa*;
6.  an interaction term coded 1=female freshman, 0 otherwise;
7.  an interaction term coded 1=female sophomore, 0 otherwise;
8.  an interaction term coded 1=female junior, 0 otherwise.

Interpret the individual dummy variables' regression coefficients as effects on predicted or mean $Y$. For example, the coefficient on the first category of *sex* (female) equals −3.840237. This tells us that the predicted drinking scale values of females are 3.840237 points lower than those of males with the same grade point average and class standing.

Regression formulations of ANOVA results help the analyst go beyond fuzzy claims that "such-and-such variables had significant effects." We can state point estimates and confidence intervals for the magnitude of effects from each category. The more complicated the ANOVA, the more helpful such specificity becomes.

## Also Type `help`

| | |
|---|---|
| `anova` | general ANOVA, ANCOVA, or regression |
| `bitest` | exact binomial probabilities test |
| `estimate` | temporarily hold results from last estimation in memory |
| `kappa` | interrater reliability |
| `ksmirnov` | Kolmogorov-Smirnov one or two-variable test |
| `kwallis` | Kruskal-Wallis $K$-sample test |
| `loneway` | large oneway ANOVA, random effects, and reliability |
| `oneway` | oneway ANOVA, multiple comparison tests |
| `predict` | ANOVA cell means, residuals, diagnostics |
| `sdtest` | test equality of standard deviations or variances |
| `serrbar` | standard error-bar chart |
| `signrank` | Wilcoxon matched-pairs signed-rank test |
| `tab4` | many oneway ANOVA, variable list by first-named variable (supplied with Student Stata) |
| `ttest` | one and two-sample $t$ tests |

# 6
# Bivariate Regression

A bivariate regression command has the form **regress Y X**, where *Y* represents the dependent variable and *X* the independent variable. (**regress** can accommodate more than one *X* variable, as shown in Chapter 7.) Like other Stata analytical commands, **regress** may be restricted to certain cases using the qualifiers **in** or **if**, or repeated for subgroups using **by**.

## The Regression Table

For illustration we turn to *stated.dta*, data on education variables for 21 U.S. states where the Scholastic Aptitude Test (SAT) is the predominant college admissions exam:

```
. describe

Contains data from c:\stustata\stated.dta
  Obs:    21 (max=  2620)               SAT Scores in 21 U.S. States
  Vars:    6 (max=    99)
Width:    13 (max=   200)
  1. state       byte   %8.0g    name
  2. SAT         int    %8.0g             average SAT scores '82
  3. educate     float  %9.0g             median ed. of pop 25+
  4. salary      int    %8.0g             average teachers' salaries
  5. expend      int    %8.0g             per pupil expenditure '82
  6. income      int    %8.0g             1984 per capita income
Sorted by:  state
```

Government and political leaders sometimes make much of state average SAT scores, depicting them as a "report card" that tells how good a job the state's schools are doing. But average scores also reflect a state's demographic makeup. For example, regressing average SAT scores on the median education level of the state's adult population:

```
. regress SAT educate

   Source |       SS       df       MS                  Number of obs =      21
----------+------------------------------               F( 1,    19) =   27.80
    Model | 12037.942       1   12037.942               Prob > F      =  0.0000
 Residual | 8228.62944     19   433.08576               R-square      =  0.5940
----------+------------------------------               Adj R-square  =  0.5726
    Total | 20266.5714     20  1013.32857               Root MSE      =  20.811


      SAT |     Coef.   Std. Err.       t     P>|t|      [95% Conf. Interval]
----------+---------------------------------------------------------------------
  educate | 146.2437   27.73881     5.272    0.000      88.18565     204.3017
    _cons | -946.0102  345.9724    -2.734    0.013      -1670.139    -221.8816
```

This regression table begins with an overall *F* test, based on the sums of squares at upper left. We test the null hypothesis that coefficients on all *X* variables in the model (here there is only one *X* variable, *educate*) equal zero. The *F* statistic, 27.8 with 1 and 19 degrees of freedom, leads easily to rejection of this hypothesis. Prob > F means the probability of a greater *F* statistic if the null hypothesis were true.

At upper right we also see the coefficient of determination, $R^2 = .5940$. Median education level explains over 59% of the variance in state SAT scores. *Adjusted $R^2$* (written $R^2_a$) takes into account the complexity of the model relative to the complexity of the data, and is often preferred for multivariate work.

The bottom half of the table gives the regression model itself. We find coefficients (slope and $Y$-intercept) in the first column, here yielding the prediction equation:

$$SAT = -946.0102 + 146.2437educate$$

The second column contains estimated standard errors of the coefficients. These are used to form both $t$ tests (columns 3–4) and confidence intervals (columns 5–6) for each regression coefficient. The $t$ statistics (coefficients divided by standard errors) test null hypotheses that corresponding population coefficients equal zero. At the $\alpha = .05$ significance level, we could reject this null hypothesis for both the coefficient on *educate* (".000", meaning $P < .0005$) and the $Y$-intercept ($P = .013$). Stata's model fitting commands (like **regress**) calculate 95% confidence intervals automatically, but we can ask for some other level of confidence. For example, to get 99% confidence intervals:

```
. regress SAT educate, level(99)
```

The term **_cons** stands for the regression constant, usually set at one. Stata automatically includes a constant unless we tell it not to. The **nocons** option causes Stata to suppress the constant, performing regression through the origin. For example:

```
. regress Y X, nocons
```

or

```
. regress Y X1 X2 X3 X4, nocons
```

In certain advanced applications, you may wish to specify your own constant. If the $X$ variables include a user-supplied constant, employ the **hascons** option instead of **nocons**:

```
. regress Y X1 X2 X3 X4, hascons
```

Using **nocons** in this situation would produce a misleading $F$ test and $R^2$. Consult the *Stata Reference Manual* (and see **help regress**) for more about **hascons**.

## Predicted Values and Residuals

After any regression, **predict** can calculate predicted values, residuals, and other case statistics. To create a new variable called *yhat*, containing predicted $Y$ values from the previous regression, type:

```
. predict yhat
. label variable yhat "predicted mean SAT score"
```

By adding the **resid** option, we create another new variable called *e*, containing residuals:

```
. predict e, resid
. label variable e "residual"
```

We might instead have obtained the same predicted $Y$ and residuals through two **generate** commands:

```
. generate yhat = _b[_cons]+_b[educate]*educate
. generate e = SAT-yhat
```

Stata remembers coefficients and other details from the most recent regression; _b[varname] holds the regression coefficient on *X* variable *varname*. _b[_cons] holds the coefficient on _cons, i.e. the *Y*-intercept. **predict** saves us the work of generating *yhat* and *e* "by hand" in this fashion.

Residuals contain information about where the model fits poorly, and so are a major focus of diagnostic analysis. Such analysis may begin with simply sorting and listing the residuals:

```
. sort e
. list state educate SAT yhat e
```

| | state | educate | SAT | yhat | e |
|---|---|---|---|---|---|
| 1. | Hawaii | 12.7 | 857 | 911.2842 | -54.28426 |
| 2. | S. Carol | 12.1 | 790 | 823.5381 | -33.53807 |
| 3. | Georgia | 12.2 | 823 | 838.1624 | -15.16244 |
| 4. | New Jers | 12.5 | 869 | 882.0355 | -13.03553 |
| 5. | Californ | 12.7 | 899 | 911.2842 | -12.28426 |
| 6. | N. Carol | 12.2 | 827 | 838.1624 | -11.16244 |
| 7. | Massachu | 12.6 | 888 | 896.6599 | -8.659899 |
| 8. | Indiana | 12.4 | 860 | 867.4112 | -7.411168 |
| 9. | Oregon | 12.7 | 908 | 911.2842 | -3.284264 |
| 10. | Connecti | 12.6 | 896 | 896.6599 | -.6598985 |
| 11. | Texas | 12.4 | 868 | 867.4112 | .5888325 |
| 12. | Florida | 12.5 | 889 | 882.0355 | 6.964467 |
| 13. | Maryland | 12.5 | 889 | 882.0355 | 6.964467 |
| 14. | Vermont | 12.6 | 904 | 896.6599 | 7.340102 |
| 15. | Maine | 12.5 | 890 | 882.0355 | 7.964467 |
| 16. | New York | 12.5 | 896 | 882.0355 | 13.96447 |
| 17. | Delaware | 12.5 | 897 | 882.0355 | 14.96447 |
| 18. | Pennsylv | 12.4 | 885 | 867.4112 | 17.58883 |
| 19. | Virginia | 12.4 | 888 | 867.4112 | 20.58883 |
| 20. | New Hamp | 12.6 | 925 | 896.6599 | 28.3401 |
| 21. | Rhode Is | 12.3 | 887 | 852.7868 | 34.2132 |

**predict** also obtains other diagnostic measures regarding the last regression. Here is a complete list of **predict** options:

| | |
|---|---|
| . **predict** *newvar* | predicted *Y* |
| . **predict** *newvar*, cooksd | Cook's *D* influence statistic |
| . **predict** *newvar*, hat | diagonal elements of hat matrix (leverage) |
| . **predict** *newvar*, index | index function (after **logit**, **probit**, or **cox**) |
| . **predict** *newvar*, resid | residuals |
| . **predict** *newvar*, rstandard | standardized residuals |
| . **predict** *newvar*, rstudent | studentized residuals |
| . **predict** *newvar*, stdf | standard error of predicted individual *Y* |
| . **predict** *newvar*, stdp | standard error of predicted mean *Y* |
| . **predict** *newvar*, stdr | standard error of residual |

When using **predict**, substitute a new variable name of your choosing for the "*newvar*" shown above. For example, to get standard errors for predicting the conditional mean of *Y*:

```
. predict SE, stdp
```

or standard errors for confidence intervals predicting individual-case *Y* values:

```
. predict SEyhat, stdf
```

The **predict**ed variables' names (such as *yhat, e, SE, SEyhat*) are arbitrary, invented by the user. Like other elements of Stata commands, options can be shortened to their first few letters—as many letters as it takes to uniquely identify the option. For example,

```
. predict e, resid
```

could be shortened to

```
. pre e, re
```

The next section shows some graphical uses for variables defined by **predict**.

## Graphing

Predicted values lie on the regression line. By plotting and connecting these values, we make the regression line visible (Figure 6.1):

```
. graph SAT yhat educate, connect(.s) symbol(Oi)
        ylabel(800,820,840,860,880,900,920)
        xlabel(12.1,12.2,12.3,12.4,12.5,12.6,12.7)
        title(Regression of State SAT Scores on Median Education)
```

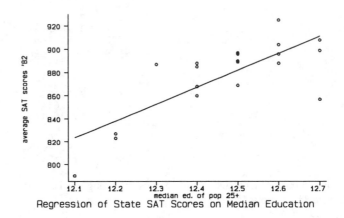

Regression of State SAT Scores on Median Education

**Figure 6.1**

**connect(.s)** tells **graph** not to connect the first-named *Y* variable (*SAT*), but to connect the second-named (*yhat*) with smooth lines. **symbol(Oi)** specifies that *SAT* should be plotted with large circles, and *yhat* invisibly—so we see only the connecting line. **connect(OT)**, for example, would have shown *yhat* values as triangles.

Residual versus predicted *Y* plots are a basic diagnostic tool. Figure 6.2 shows an example with marginal boxplots and a horizontal line at the mean residual, *e* = 0:

```
. graph e yhat, twoway box yline(0) ylabel xlabel title(Residual vs.
        Predicted Values Plot:  State SAT Scores)
```

By adding **oneway**, we could draw oneway scatterplots as well as boxplots in the margins. The **rvfplot** command draws graphs like Figure 6.2 automatically.

Figure 6.2 hints of a curvilinear pattern in the residuals. Perhaps the straight-line model of Figure 6.1 is unrealistic, and we should consider fitting a curve instead. Variable transformation, described later in this chapter, provides the simplest way to do this.

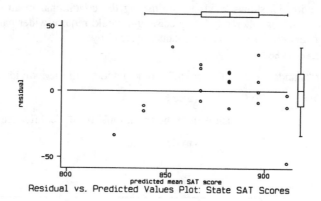

Residual vs. Predicted Values Plot: State SAT Scores

**Figure 6.2**

The command

```
. predict SE, stdp
```

creates a new variable (arbitrarily) called *SE*, containing standard errors for predicting the conditional mean. To calculate endpoints of a 95% confidence interval, we use these standard errors and an appropriate *t* value:

```
. generate hi = yhat+2.093*SE
. label variable hi "upper 95% confidence limit"
. generate lo = yhat-2.093*SE
. label variable lo "lower 95% confidence limit"
```

Confidence bands for regression predictions exhibit an hourglass shape, narrowest at the mean of *X* (Figure 6.3):

```
. graph SAT yhat hi lo educate, connect(.sss) symbol(Oiii) ylabel
        xlabel title(95% Confidence Bands for Predicted Mean SAT Scores)
```

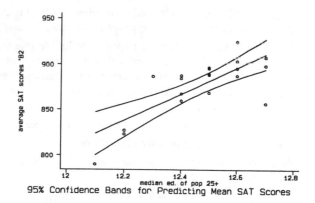

95% Confidence Bands for Predicting Mean SAT Scores

**Figure 6.3**

Note that Figure 6.3 shows bands for estimating the conditional mean of $Y$. Predicting individual-case $Y$ values is inherently less precise and would require wider bands. The correct standard error for that purpose should be obtained instead by:

```
. predict SEyhat, stdf
```

**stdf** stands for "standard error of forecast." Some authors call the individual-case interval

$$yhat \pm t \times SEyhat \qquad \textbf{(stdf)}$$

a "prediction interval" to distinguish it from the conditional-mean "confidence interval:"

$$yhat \pm t \times SE \qquad \textbf{(stdp)}$$

## Correlation Matrices

**correlate** obtains Pearson product-moment correlations between variables:

```
. correlate SAT educate
(obs=21)
         |     SAT  educate
---------+------------------
     SAT|  1.0000
 educate|  0.7707   1.0000
```

To see a correlation matrix, type **correlate** followed by the variable list:

```
. correlate educate salary expend SAT income
(obs=21)
         | educate   salary   expend      SAT   income
---------+---------------------------------------------
 educate|  1.0000
  salary|  0.4844   1.0000
  expend|  0.5202   0.7622   1.0000
     SAT|  0.7707   0.3513   0.5559   1.0000
  income|  0.4850   0.6032   0.6439   0.4587   1.0000
```

Variance-covariance matrices are obtained with the **covariance** option, as in:

```
. correlate educate salary expend SAT income, covariance
```

Typing **correlate, _coef** obtains the estimated correlation matrix of coefficients from the last regression; for their variance-covariance matrix type **correlate, _coef covariance**.

Unlike some statistical packages, Stata declines to print $P$-values testing the significance of each correlation in the matrix. Correlation matrices with $P$-values tempt unwary analysts to commit multiple-comparison fallacies—scanning the matrix for "significant" correlations, then entering those variables in a model and (circularly) discovering their "significance." Determined Stata users who need tests for specific correlations can either:

1.   refer to the equivalent $F$ or $t$ tests in a bivariate regression (**regress**);
2.   write their own program, as illustrated on pages 24–25; or
3.   with small samples, obtain Kendall's $\tau$ rank correlation (slow to compute). For example: **ktau expend salary**

**ktau** does not require **regress**'s constant-variance assumption, nor does it assume normality in small samples. **ktau** (like the Spearman rank correlation, **spearman**) also has better outlier resistance than **regress** or **correlate**.

Approximate nonparametric versions of many statistical procedures can be produced by first transforming the variables to ranks, using **egen**, then performing parametric analysis (**regress**,

**correlate**, etc.) as usual.   The next section describes another way to accommodate problematic variables within the general framework of regression analysis.

## Regression with Transformed Variables

Power transformations (Chapter 2) play an important role in regression analysis, where they may be called upon to linearize relations, stabilize variance, normalize distributions, and/or pull in outliers.   Sometimes a transformation needed for one purpose will simplify the analysis in other respects as well.

Dataset *nevada.dta* contains information about highway fatality rates in 17 Nevada counties (Baker, Whitfield, and O'Neill 1987):

```
. describe
```

```
Contains data from c:\stustata\nevada.dta
  Obs:    17 (max=  2620)              Baker et al. (1987)
  Vars:    3 (max=    99)
Width:     9 (max=   200)
   1. county      byte   %8.0g   cname   Nevada county
   2. fatal       float  %9.0g           fatal crashes/100K residents
   3. density     float  %9.0g           1980 population/square mile
Sorted by:
```

```
. list county density fatal
```

```
         county    density     fatal
  1.    Churchil        2.8    69.459
  2.       Clark       58.8    27.353
  3.     Douglas       27.4    53.207
  4.        Elko          1    94.582
  5.    Esmerald         .2   557.701
  6.      Eureka         .3   389.538
  7.    Humboldt          1   173.133
  8.      Lander         .7   130.847
  9.     Lincoln         .4   133.976
 10.        Lyon        6.8     95.63
 11.     Mineral        1.7     64.34
 12.         Nye         .5    136.31
 13.    Pershing         .6   185.837
 14.      Storey        5.7   110.889
 15.      Washoe       30.7     17.56
 16.    White_Pi         .9    93.874
 17.    Carson_C      219.9    26.024
```

Is fatality rate related to population density?   Linear regression finds a nonsignificant negative slope:

```
. regress fatal density
```

| Source | SS | df | MS |
|---|---|---|---|
| Model | 31520.4678 | 1 | 31520.4678 |
| Residual | 274443.918 | 15 | 18296.2612 |
| Total | 305964.386 | 16 | 19122.7741 |

| | |
|---|---|
| Number of obs = | 17 |
| F( 1, 15) = | 1.72 |
| Prob > F = | 0.2091 |
| R-square = | 0.1030 |
| Adj R-square = | 0.0432 |
| Root MSE = | 135.26 |

| fatal | Coef. | Std. Err. | t | P>\|t\| | [95% Conf. Interval] |
|---|---|---|---|---|---|
| density | -.8277648 | .6306549 | -1.313 | 0.209 | -2.171974    .5164444 |
| _cons | 156.3387 | 35.41206 | 4.415 | 0.001 | 80.85973    231.8178 |

We follow the usual steps to graph this regression (Figure 6.4):

```
. predict yhat1
. label variable yhat1 "predicted fatality rate"
. graph fatal yhat1 density, connect(.s) symbol(Oi) ylabel xlabel
```

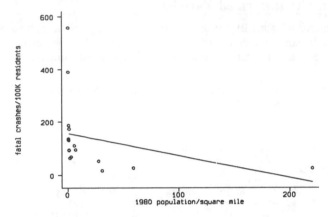

**Figure 6.4**

The fit in Figure 6.4 looks poor. Most of the data points cluster at lower left, because both variables are positively skewed. The regression line goes nowhere near the data, and unrealistically predicts negative fatality rates at higher densities. The highest-density county (Carson) looks influential, and residuals cannot be close to normal. In short, Figure 6.4 shows that almost everything is wrong with this regression.

Taking logarithms of both variables works wonders here.

```
. generate ldense = ln(density)/ln(10)
. label variable ldense "logarithm of population density"
. generate lfatal = ln(fatal)/ln(10)
. label variable lfatal "logarithm of fatality rate"
. regress lfatal ldense
```

| Source | SS | df | MS | | Number of obs = | 17 |
|--------|----|----|----|----|-----------------|----|
| Model | 1.85404704 | 1 | 1.85404704 | | F( 1, 15) = | 46.93 |
| Residual | .592571009 | 15 | .039504734 | | Prob > F = | 0.0000 |
| | | | | | R-square = | 0.7578 |
| | | | | | Adj R-square = | 0.7417 |
| Total | 2.44661805 | 16 | .152913628 | | Root MSE = | .19876 |

| lfatal | Coef. | Std. Err. | t | P>|t| | [95% Conf. Interval] |
|--------|-------|-----------|---|-------|----------------------|
| ldense | -.3828331 | .0558822 | -6.851 | 0.000 | -.5019433   -.263723 |
| _cons | 2.137288 | .0533584 | 40.055 | 0.000 | 2.023557   2.251019 |

```
. predict yhat2
. label variable yhat2 "predicted log fatality rate"
```

A graph of the regression of log fatality rate (*lfatal*) on log density (*ldense*) shows none of the statistical problems seen earlier in Figure 6.4 (Figure 6.5):

```
. graph lfatal yhat2 ldense, connect(.s) symbol(Oi) ylabel xlabel
```

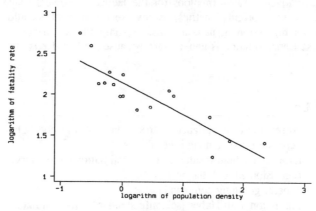

**Figure 6.5**

To see what the linear log-log relation of Figure 6.5 implies, we apply an inverse transformation to the predicted $Y$ values. This returns predicted $Y$ to the original units of $Y$. Since we transformed fatality rate by taking base 10 logs, the appropriate inverse transformation involves antilogs: raising 10 to the power of predicted log fatalities.

```
. generate yhat3 = 10^yhat2
. label variable yhat3 "predicted fatality rate"
```

*yhat2* was measured in log(fatalities/100,000 people); *yhat3* = 10^*yhat2* is measured simply in fatalities/100,000 people. Graphing *yhat3* values against *density* (not log density) reveals a regression curve (Figure 6.6):

```
. graph fatal yhat3 density, connect(.s) symbol(Oi) ylabel xlabel
```

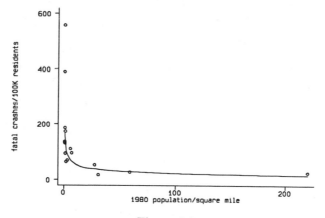

**Figure 6.6**

The curve in Figure 6.6 fits the data better than the straight line in Figure 6.4 does and no longer yields negative predicted fatality rates. It depicts the rates dropping steeply as *density* increases from very low to low, then almost leveling off.

Chapter 2 gave inverse transformations for the ladder of powers (page 28). Returning predicted $Y$ values to their original metrics, by inverse transformation, allows us to graph a transformed-variables regression in natural units. If only the $X$ variables (but not $Y$) were transformed, inverse transformation is unnecessary because predicted $Y$ is already in the natural units of $Y$.

## Also Type help

| | |
|---|---|
| correlate | correlation or covariance matrix of variables or coefficients |
| egen | generate ranks, among other things |
| estimate | temporarily hold results from last estimation in memory |
| fit | regression fit and diagnostics |
| graph | general graphing command |
| hreg | regression with Huber jackknife standard error estimates |
| predict | predicted values, residuals, and diagnostic statistics |
| qreg | quantile regression |
| regress | regression analysis (OLS, WLS, 2SLS, IV) |
| rreg | robust regression |
| spearman | Spearman or Kendall rank correlation coefficients |

# 7
# Multiple Regression

Chapter 6 illustrated bivariate regression, obtained by commands of the form

`. regress Y X`

The **regress** command handles multiple regression too, with (almost) any number of *X* variables:

`. regress Y X1 X2 X3 X4 ...`

Regression adapts to a wide variety of research problems. This chapter introduces Stata's basic regression capabilities.

## Basic Multiple Regression

For illustration, we will use data on grades and teaching evaluations in 118 college courses:

`. use c:\stustata\teval`
(Teaching Evaluations & Grades)

`. describe`

```
Contains data from c:\stustata\teval.dta
  Obs:   118 (max=  2620)              Teaching Evaluations & Grades
  Vars:    4 (max=    99)
Width:     5 (max=   200)
   1. evals          byte   %8.0g          % high evaluations rec'd
   2. grades         byte   %8.0g          % As and Bs given
   3. size           int    %8.0g          class size
   4. fac            byte   %8.0g   faclbl instructor's status
Sorted by:  fac
```

Forty-seven of the 118 classes were taught by part-time instructors or teaching assistants; regular tenure-track faculty taught the other 71 classes. The variable *fac*, coded 1 for faculty-taught classes and 0 for others, indicates instructor status.

We will investigate the effects of grades and class size on teaching evaluations, first considering only those classes taught by nonfaculty. The qualifier **if fac==0** (note double equals) restricts this regression to nonfaculty classes:

`. regress evals grades size if fac==0`

```
  Source |       SS       df       MS                  Number of obs =      47
---------+------------------------------              F(  2,    44) =    5.41
   Model |  5721.65592     2  2860.82796              Prob > F      =  0.0079
Residual |  23270.9824    44  528.885963              R-square      =  0.1973
---------+------------------------------              Adj R-square  =  0.1609
   Total |  28992.6383    46  630.274746              Root MSE      =  22.998


------------------------------------------------------------------------------
   evals |      Coef.   Std. Err.       t    P>|t|       [95% Conf. Interval]
---------+--------------------------------------------------------------------
  grades |   .7323049   .2878445      2.544   0.015       .1521925    1.312417
    size |  -.3031118   .1308428     -2.317   0.025      -.5668081   -.0394156
   _cons |   11.56337   18.85146      0.613   0.543      -26.42926    49.55599
------------------------------------------------------------------------------
```

We obtain the following regression equation for predicting the percentage of high teaching evaluations (*evals*) in nonfaculty courses:

$$evals = 11.56337 + .7323049grades - .3031118size \qquad [7.1]$$

The regression coefficients on both *grades* and *size* are statistically significant at the .05 level ($P = .015$ and $P = .025$, respectively). These first-order partial regression coefficients are interpreted as follows:

$b_1 = .7323049$: Predicted high evaluations increase by about .73 percentage points with each 1-percentage point increase in high grades if class size stays constant.

$b_2 = -.3031118$: Predicted high evaluations decrease by about .30 percentage points with each 1-student increase in class size if the percentage of high grades stays constant.

*grades* and *size* together explain about 19.73% of the variance in evaluations or 16.09% adjusting for degrees of freedom ($R_a^2 = .1609$).

Using the qualifier **if fac==1**, we can repeat this analysis for the other part of the sample, the 71 classes taught by regular faculty members:

. **regress evals grades size if fac==1**

```
  Source |       SS       df       MS                    Number of obs =      71
---------+------------------------------                 F(  2,    68) =    4.55
   Model |  4196.63483     2   2098.31741                Prob > F      =  0.0140
Residual |  31374.5483    68   461.390416                R-square      =  0.1180
---------+------------------------------                 Adj R-square  =  0.0920
   Total |  35571.1831    70   508.159759                Root MSE      =   21.48


--------------------------------------------------------------------------------
   evals |     Coef.    Std. Err.       t     P>|t|      [95% Conf. Interval]
---------+----------------------------------------------------------------------
  grades |   .5119427    .1700349     3.011   0.004      .1726433     .8512421
    size |   .0126773     .093712     0.135   0.893     -.174322     .1996766
   _cons |   24.84383    11.11152     2.236   0.029      2.671133    47.01652
--------------------------------------------------------------------------------
```

The regression equation for faculty-taught classes appears notably different from [7.1]:

$$evals = 24.84383 + .5119427grades + .0126773size \qquad [7.2]$$

Grades still have a significant ($P = .004$), positive effect. The coefficient on class size, however, is nonsignificant ($P = .893$) and near zero (.0126773, compared with $-.3031118$ for nonfaculty classes). Among faculty-taught classes, grades and size explain only about 9% of the variance in evaluations ($R_a^2 = .092$), compared with 16% among nonfaculty classes. The following sections examine ways to make faculty/nonfaculty differences an explicit part of the model.

With any regression, we can get standardized regression coefficients or "beta weights" in addition to the usual unstandardized coefficients, by including the **beta** option (output not shown):

. **regress evals grades size, beta**

## Dummy Variables

"Dummy variable" refers to any two-category variable coded 0 and 1. In *teval.dta*, *fac* is a dummy variable. Regression ordinarily requires a measurement $Y$ variable, but permits any mixture of measurement and dummy $X$ variables. For a simple example, regress *evals* on *fac*:

```
. regress evals fac

    Source |       SS        df       MS                    Number of obs =      118
-----------+------------------------------                  F(  1,   116) =     4.60
     Model |  2562.95826      1   2562.95826                Prob > F      =   0.0340
  Residual |  64563.8214    116   556.584667                R-square      =   0.0382
-----------+------------------------------                  Adj R-square  =   0.0299
     Total |  67126.7797    117   573.733159                Root MSE      =   23.592


------------------------------------------------------------------------------
     evals |     Coef.    Std. Err.        t     P>|t|     [95% Conf. Interval]
-----------+------------------------------------------------------------------
       fac |  9.519928    4.436378      2.146    0.034    .7331233    18.30673
     _cons |  43.17021    3.441254     12.545    0.000    36.35438    49.98605
------------------------------------------------------------------------------
```

Regression with a dummy as the only $X$ variable performs a difference-of-means test, equivalent to tests we could obtain through a two-sample $t$ test or ANOVA (Chapter 4). To understand any dummy variable regression, write out the equation and then substitute 0's and 1's. For example:

$$evals = 43.17021 + 9.519928 fac \qquad [7.3]$$

When *fac* equals 0 (nonfaculty classes), [7.3] simplifies to:

$$evals = 43.17021 + 9.519928 \times 0$$
$$= 43.17021$$

When *fac* equals 1 (faculty classes), [7.3] yields:

$$evals = 43.17021 + 9.519928 \times 1$$
$$= 52.690138$$

These two predicted values equal the subgroup means (within roundoff error):

```
. tab fac, summ(evals)

instructor's| Summary of % high evaluations rec'd
     status|        Mean    Std. Dev.        Freq.
-----------+------------------------------------
     T. A. |   43.170213   25.105273           47
    faculty |   52.690141   22.542399           71
-----------+------------------------------------
     Total |   48.898305   23.952728          118
```

In regression with a single dummy $X$ variable, the coefficient on $X$ equals the difference in subgroup means; a $t$ test of whether this coefficient equals zero amounts to a test for a significant difference between means. In the example above, the difference is significant ($P = .035$); faculty tend to get better evaluations than nonfaculty.

When a dummy variable is included with one or more measurement $X$ variables, it plays the role of an "intercept dummy variable." For example:

```
. regress evals grades size fac
```

```
  Source |       SS        df       MS                    Number of obs =      118
---------+------------------------------                  F(  3,   114) =     6.93
   Model | 10357.4694      3   3452.48981                 Prob > F      =   0.0003
Residual | 56769.3102    114   497.976406                 R-square      =   0.1543
---------+------------------------------                  Adj R-square  =   0.1320
   Total | 67126.7797    117   573.733159                 Root MSE      =   22.315


------------------------------------------------------------------------------
   evals |     Coef.    Std. Err.       t     P>|t|      [95% Conf. Interval]
---------+--------------------------------------------------------------------
  grades |   .5477258    .148646      3.685   0.000      .2532592    .8421924
    size |  -.0999158    .0769265    -1.299   0.197     -.2523066    .052475
     fac |   15.57809    4.466984     3.487   0.001      6.729032   24.42715
   _cons |   13.51118    10.66279     1.267   0.208     -7.611718   34.63408
------------------------------------------------------------------------------
```

Again, to understand these results we can write out the equation substituting 0 and 1:

$$evals = 13.51118 + .5477258grades - .0999158size + 15.57809fac \qquad [7.4]$$

For nonfaculty classes, [7.4] becomes:

$$evals = 13.51118 + .5477258grades - .0999158size + 15.57809 \times 0$$
$$= 13.51118 + .5477258grades - .0999158size \qquad [7.4a]$$

For faculty classes:

$$evals = 13.51118 + .5477258grades - .0999158size + 15.57809 \times 1$$
$$= 29.08927 + .5477258grades - .0999158size \qquad [7.4b]$$

Equations [7.4a] and [7.4b] differ only in their $Y$-intercepts, hence the term "intercept dummy variable." The $t$ test of the coefficient on *fac* ($t = 3.487$, $P = .001$) indicates that these intercepts are significantly different.

Separate regressions for nonfaculty and faculty classes (equations [7.1] and [7.2]) earlier found different intercepts and slopes. We allow for slope differences by including "slope dummy variables," formed by multiplying a dummy times a measurement variable:

```
. generate facgrade = fac*grades
. generate facsize = fac*size
```

To see what this does, we can look at the first six cases. `nolabel` suppresses value labels:

```
. list grades size fac facgrade facsize in 1/6, nolabel

       grades      size       fac   facgrade    facsize
  1.       75        96         1         75         96
  2.       45        60         1         45         60
  3.       41        86         1         41         86
  4.       53        58         0          0          0
  5.       66        29         0          0          0
  6.       74        43         1         74         43
```

Regression with the full set of intercept and slope dummy variables (a slope dummy for each measurement $X$ variable) can duplicate the results from separate-sample regression:

```
. regress evals grades size fac facgrade facsize
```

```
  Source |      SS        df       MS                  Number of obs =      118
---------+-----------------------------               F( 5,    112) =     5.12
   Model |   12481.249     5   2496.2498               Prob > F      =   0.0003
Residual |  54645.5307   112  487.906524               R-square      =   0.1859
---------+-----------------------------               Adj R-square  =   0.1496
   Total |  67126.7797   117  573.733159               Root MSE      =   22.089


----------------------------------------------------------------------------
   evals |     Coef.    Std. Err.        t      P>|t|     [95% Conf. Interval]
---------+------------------------------------------------------------------
  grades |   .7323049   .2764682      2.649    0.009     .1845186    1.280091
    size |  -.3031118   .1256716     -2.412    0.017    -.5521139   -.0541098
     fac |   13.28046   21.41036      0.620    0.536    -29.14142    55.70234
 facgrade|  -.2203622   .3271209     -0.674    0.502    -.8685104     .427786
 facsize |   .3157892   .1583666      1.994    0.049     .0020061    .6295722
   _cons |   11.56337   18.10641      0.639    0.524    -24.31216     47.4389
----------------------------------------------------------------------------
```

The regression equation now is:

*evals*     = 11.56337 + .7323049*grades* − .3031118*size* + 13.28046*fac* −.2203622*facgrade*
            + .3157892*facsize*                                                   [7.5]

which simplifies for *fac*=0 classes to:

*evals*     = 11.56337 + .7323049*grades* − .3031118*size* + 13.28046×0 −.2203622×0
            + .3157892×0

            = 11.56337 + .7323049*grades* − .3031118*size*                        [7.5a]

Equation [7.5a], obtained by substituting *fac* = 0 into the slope and intercept dummy variable regression, is identical to equation [7.1], obtained by restricting the regression to the *fac* = 0 subsample. Similarly, when *fac* = 1 equation [7.5] simplifies to:

*evals*     = 11.56337 + .7323049*grades* − .3031118*size* + 13.28046×1 −.2203622*grades*
            + .3157892*size*

            = 24.84383 + .5119427*grades* + .0126774*size*                        [7.5b]

Equation [7.5b] is identical to the separate-sample model, equation [7.2].

Regression with intercept and slope dummy variables does not just duplicate separate-sample regressions; it also provides tests of where they significantly differ. The coefficient on *fac* (13.28046) equals the difference in intercepts between [7.5a] and [7.5b]. Its *t* statistic (*t* = .620, *P* = .536) tells us that this difference is not statistically significant. Similarly, the coefficient on *facgrade* equals the difference in *grades'* coefficients (also not significant, *P* = .502), comparing [7.5a] with [7.5b]. Only the coefficient on *size* differs significantly (*P* = .049) between nonfaculty and faculty classes. The amount of this difference equals the coefficient on *facsize*, .3157892.

## Testing Hypotheses

With every regression, Stata prints two kinds of hypothesis tests:

1.  Overall *F* test: The *F* statistic at upper right tests the null hypothesis that in the population, coefficients on all the *X* variables equal zero.

2.  Individual *t* tests: The third and fourth columns of the regression table contain *t* tests for each individual regression coefficient. These test the null hypothesis that in the population, the coefficient on that particular *X* variable equals zero.

The individual $t$ tests are always two-sided. For a one-sided test, divide the output's $P$-value in half.

In addition to these standard $F$ and $t$ tests, Stata can perform $F$ tests of user-specified hypotheses. The **test** command refers back to the most recent model fitting (**regress**, **anova**, etc.) command. For example, test whether all the coefficients on any subset of $X$ variables in the last analysis equal zero, type **test** followed by the variables' names:

```
. test fac facgrade

( 1)   fac = 0.0
( 2)   facgrade = 0.0

       F(  2,   112) =     0.23
            Prob > F =     0.7972
```

Above we see a test for the null hypothesis that coefficients on both *fac* and *facgrade* (from the last regression) equal zero. We cannot reject this null hypothesis ($P = .7972$). Tests on subsets of $X$ variables are useful when we have several conceptually related predictors or when individual coefficient estimates are unreliable due to multicollinearity.

**test** could duplicate the overall $F$ test:

```
. test grades size fac facgrade facsize
```

or the individual-coefficient tests:

```
. test grades
. test size
. test fac
```

and so forth. Applications of **test** more useful in advanced work include:

1.  **test** whether a coefficient equals a specified constant. For example, to test the null hypothesis that the coefficient on *grades* equals 1, instead of the usual null hypothesis that it equals 0:

    ```
    . test grades = 1
    ```

2.  **test** whether two or more coefficients are equal. For example:

    ```
    . test grades = size
    ```

    evaluates the null hypothesis $H_0 : \beta_1 = \beta_2$ .

**test** understands some algebraic expressions, so we can even ask for tests like:

```
. test grades=(fac+facsize)/100
```

Type **help test** for more information and examples.

## Stepwise Regression

Equation [7.5] includes five predictors, two of them not distinguishable from zero at the .05 significance level. We could improve the model's parsimony by dropping nonsignificant predictors. This is best done one variable at a time, with the model re-estimated after each elimination to see what should go next.

Stepwise regression automates this process of backward elimination. We can list all our $X$ variables and a minimum $F$ statistic value required to retain them. With 1 and 120 degrees of freedom, $F = 3.92$ is the cutoff for .05 level significance. Adopting 3.92 as our minimum $F$-to-stay therefore asks that only variables with coefficients significant at .05 be kept in the model:

```
. stepwise evals grades size fac facgrade facsize, backward fstay(3.92)
```

```
Dropping: fac       F=      0.3847
Dropping: facgrade  F=      0.0698
```

```
                                                           (stepwise)
  Source |     SS        df       MS          Number of obs =     118
---------+------------------------------      F( 3,   114) =      8.49
   Model | 12259.6387     3   4086.54622      Prob > F      =   0.0000
Residual | 54867.141    114   481.29071       R-square      =   0.1826
---------+------------------------------      Adj R-square  =   0.1611
   Total | 67126.7797   117   573.733159      Root MSE      =   21.938
```

```
-----------------------------------------------------------------------
   evals |    Coef.    Std. Err.      t     P>|t|    [95% Conf. Interval]
---------+-------------------------------------------------------------
  grades |   .5746733   .1463444     3.927   0.000    .2847662   .8645804
    size |  -.2980006   .0937097    -3.180   0.002   -.4836387  -.1123625
 facsize |   .3153688   .0775545     4.066   0.000    .1617339   .4690038
   _cons |  21.24491    9.572378     2.219   0.028    2.282104  40.20772
-----------------------------------------------------------------------
```

**stepwise** dropped first *fac*, then *facgrade*, settling on the final model:

$$evals = 21.24491 + .5746733 grades - .2980006 size + .3153688 facsize \qquad [7.6]$$

Because [7.6] is simpler than [7.5] but fits nearly as well, its $R_a^2$ is actually higher.

Among nonfaculty classes, [7.6] implies:

$$evals = 21.24491 + .5746733 grades - .2980006 size + .3153688 \times 0$$
$$= 21.24491 + .5746733 grades - .2980006 size \qquad [7.6a]$$

For faculty classes, [7.6] becomes:

$$evals = 21.24491 + .5746733 grades - .2980006 size + .3153688 size$$
$$= 21.24491 + .5746733 grades + .0173682 size \qquad [7.6b]$$

Our final model ([7.6]) could be summarized as follows:

> Among all classes, grades affect predicted evaluations. Each 1-point increase in the percentage of high grades raises the predicted percentage of high evaluations about .57 points. Among classes taught by nonfaculty, larger classes tend to get worse evaluations. The percentage of high evaluations declines about .3 points with each additional student. Among classes taught by faculty, the number of students has little impact on evaluations.

The general syntax of **stepwise** resembles that of **regress**:

```
. stepwise Y X1 X2 X3 X4 ...
```

**stepwise** offers four main options:

1. forward inclusion—start with no predictors in the equation. Add predictor variables in order of their contribution to $R^2$. Stop when no left-out variables yield $F$ statistics above the cutoff specified by **fenter**. For example:

```
. stepwise evals grades size fac facgrade facsize, forward fenter(3.92)
```

2. backward elimination—start with all possible predictors in the equation. Drop the least-valuable predictors one by one. Stop when all kept-in predictors have $F$ statistics above the cutoff specified by **fstay**. For example (demonstrated at top of this page):

```
. stepwise evals grades size fac facgrade facsize, backward fstay(3.92)
```

3. forward stepwise solution—a modification of forward inclusion in which all the brought-in variables are reexamined at each step to see whether any should be dropped again. Specify both `fstay` and `fenter`. For example:

```
. stepwise evals grades size fac facgrade facsize, forward stepwise
        fstay(3.92) fenter(3.92)
```

4. backward stepwise solution—a modification of backward elimination in which all the dropped-out variables are reexamined at each step to see whether any should be brought back in. Specify both `fstay` and `fenter`. For example:

```
. stepwise evals grades size fac facgrade facsize, backward stepwise
        fstay(3.92) fenter(3.92)
```

Different stepwise approaches often reach different conclusions from the same data. Furthermore, stepwise results tend to be sample-specific. Although stepwise procedures appeal to novices because they automate the work of selecting variables, these deceptively simple methods have numerous pitfalls and are better left to experienced analysts.

## Conditional Effect Plots

After performing any regression, Stata retains the estimated coefficients in memory as system variables denoted _b[*varname*]. For example, if we regressed

```
. regress evals grades size facsize
```

we could obtain predicted values in the usual fashion by

```
. predict yhat
```

or, alternatively, calculate the same predicted values "by hand":

```
. generate yhat = _b[_cons] + _b[grades]*grades + _b[size]*size
        + _b[facsize]*facsize
```

The second approach permits construction of conditional effect plots, which can visualize $Y-X_k$ relationships at any given level of the other $X$ variables.

The following equation describes the relation between predicted evaluations and grades, when *size* equals its mean (51.73729 students) and *fac* equals 0 (nonfaculty instructor):

```
. generate yhat0 = _b[_cons] + _b[grades]*grades + _b[size]*51.73729 if fac==0
. label variable yhat0 "Nonfaculty"
```

The qualifier `if fac==0` tells Stata to generate *yhat0* only for *fac* = 0 classes, but otherwise leave *yhat0* missing. Similarly (and using `<PgUp>`), we can get predicted values based on *size* = 51.73729 and *fac* = 1:

```
. generate yhat1 = _b[_cons] + _b[grades]*grades + _b[size]*51.73729
        + _b[facsize]*51.73729 if fac==1
. label variable yhat1 "Faculty"
```

To see these predictions graphically (Figure 7.1):

```
. graph yhat0 yhat1 grades, connect(ss) symbol(po) ylabel xlabel
        l1(% High Evaluations) title(Conditional effect plot,
        class size at mean)
```

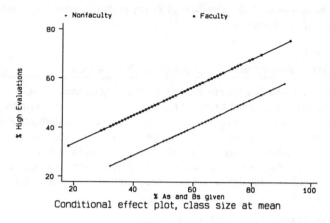

Conditional effect plot, class size at mean

**Figure 7.1**

Figure 7.1 shows that (according to our model, [7.6]) evaluations of both faculty and nonfaculty instructors improve as they give better grades. At any given level of grading, faculty instructors tend to get better evaluations. Figure 7.1 holds class size constant at its mean, but we would find similar parallel lines if we held size constant at any other level.

To visualize the effect of class size, we prepare a second conditional effect plot, this time holding *grades* constant at its mean (Figure 7.2):

```
. generate yhat2 = _b[_cons] + _b[grades]*56.9322 + _b[size]*size
            + _b[facsize]*facsize
. sort fac
. graph yhat2 size, connect(s) symbol(i) by(fac) ylabel xlabel
        ll(% High Evaluations) title(Conditional effect plot, grades at mean)
```

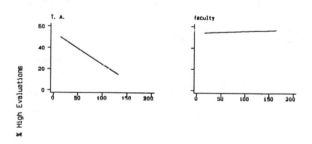

class size
Conditional effect plot, grades at mean

**Figure 7.2**

Figure 7.2 uses the **by(fac)** option to draw separate graphs for nonfaculty (T.A.) and faculty classes, emphasizing the different effects of class size. Among nonfaculty classes, evaluations worsen as class size increases. Among faculty classes, evaluations stay about the same regardless of class size. Figure 7.2 thus graphically displays the meaning of the slope dummy variable in equation [7.6]. Conditional effect plots are especially useful in picturing interactions and other nonlinear effects.

## Weighted Least Squares and Instrumental Variables

**regress** offers several further options of interest to advanced users. Weighted least squares (WLS) has many applications, including robust regression, replicated observations, and correcting for heteroscedasticity. We often suspect heteroscedasticity (nonconstant error variances) with aggregate data, when the $Y$ variable is an average based on unequal numbers of individual observations per case. Assuming that the individual observations have equal variances, we could correct for the aggregate-data heteroscedasticity by employing analytical weights (**aweights**) equal to the numbers of individual observations. For example, weighting a teaching evaluations regression by class size:

```
. regress evals grades fac [aweight=size]
(sum of wgt is  6.1050e+003)
```

| Source | SS | df | MS | | Number of obs = | 118 |
|--------|-----|-----|-----|---|---|---|
| Model | 9033.43471 | 2 | 4516.71736 | | F( 2, 115) = | 9.50 |
| Residual | 54673.5075 | 115 | 475.421805 | | Prob > F = | 0.0002 |
| | | | | | R-square = | 0.1418 |
| | | | | | Adj R-square = | 0.1269 |
| Total | 63706.9422 | 117 | 544.50378 | | Root MSE = | 21.804 |

| evals | Coef. | Std. Err. | t | P>\|t\| | [95% Conf. Interval] | |
|-------|-------|-----------|---|---------|---------------------|---|
| grades | .4436826 | .1467716 | 3.023 | 0.003 | .1529563 | .734409 |
| fac | 18.03529 | 4.477744 | 4.028 | 0.000 | 9.16574 | 26.90484 |
| _cons | 11.24169 | 9.88151 | 1.138 | 0.258 | -8.331677 | 30.81506 |

If we instead specified frequency weights (**fweight**), Stata would have incorrectly assumed that our data from 118 classes actually represented data on 6,105 individual, independently sampled students:

```
. regress evals grades fac [fweight=size]
```

| Source | SS | df | MS | | Number of obs = | 6105 |
|--------|-----|-----|-----|---|---|---|
| Model | 467365.415 | 2 | 233682.707 | | F( 2, 6102) = | 504.10 |
| Residual | 2828659.01 | 6102 | 463.562604 | | Prob > F = | 0.0000 |
| | | | | | R-square = | 0.1418 |
| | | | | | Adj R-square = | 0.1415 |
| Total | 3296024.43 | 6104 | 539.977789 | | Root MSE = | 21.531 |

| evals | Coef. | Std. Err. | t | P>\|t\| | [95% Conf. Interval] | |
|-------|-------|-----------|---|---------|---------------------|---|
| grades | .4436826 | .0201491 | 22.020 | 0.000 | .4041834 | .4831819 |
| fac | 18.03529 | .6147126 | 29.339 | 0.000 | 16.83024 | 19.24034 |
| _cons | 11.24169 | 1.356551 | 8.287 | 0.000 | 8.582374 | 13.90101 |

Notice that the coefficient estimates under either weighting are the same. Standard errors and tests come out quite differently, however. The **fweight** version, treating *size* values as counts of replications, wrongly believes we have a huge dataset here. It therefore calculates much smaller standard errors and narrower confidence intervals, reflecting the greater precision of estimates based on "$n = 6,105$" independent cases.

The **regress** command may also specify one or more instrumental variables, for example to obtain estimates for a two-equation model in which *Y1* and *Y2* influence each other:

$$Y1 = \beta_0 + \beta_1 Y2 + \beta_2 X1 + \varepsilon_1 \qquad\qquad [7.7]$$

$$Y2 = \beta_3 + \beta_4 Y1 + \beta_5 X2 + \varepsilon_2 \qquad\qquad [7.8]$$

Ordinary least squares assumes that the errors ($\varepsilon$) are independent of other predictors. But if, for instance, *Y2* is a function of *Y1*, then $\varepsilon_1$ must correlate with *Y2*. We can work around such problems of mutual influence with an instrumental variables (IV) regression. Variables exogenous to both equations appear in parentheses, and we estimate values of $\beta_0 - \beta_5$ as coefficients in these two IV regressions:

```
. regress Y1 Y2 X1 (X1 X2)
. regress Y2 Y1 X2 (X1 X2)
```

The two commands above achieve the same result as the final steps of the following two-stage least squares (2SLS) analysis:

```
. regress Y1 X1 X2
. predict Y1hat
. regress Y2 X1 X2
. predict Y2hat
. regress Y1 Y2hat X1
. regress Y2 Y1hat X2
```

See Johnston (1984) or Hanushek and Jackson (1977) for more about instrumental variables and 2SLS.


## Also Type help

| | |
|---|---|
| **anova** | analysis of variance and covariance |
| **correlate** | correlation and covariance matrices for variables and coefficients |
| **hreg** | regression with Huber jackknife standard errors |
| **impute** | replace missing values by best-subsets regression predictions |
| **pcorr** | partial correlation coefficients |
| **predict** | predicted values, residuals, or diagnostic statistics |
| **regress** | multiple regression |
| **stepwise** | stepwise regression |
| **test** | hypothesis tests |
| **weight** | **regress** allows analytical, frequency, or importance case weights |

# 8
# Regression Diagnostics

Do we have reason to distrust the analysis? Can we find ways to improve it? Careful diagnostic work, checking for assumption violations or other statistical problems, forms an important step in modern data analysis. We perform the initial regression, ANOVA, or whatever, but then look closely at our results for any signs of trouble. Some of the general statistical and graphing methods introduced in earlier chapters adapt well to diagnostic analysis. Stata also provides many specialized statistics designed for trouble-shooting.

## Correlation and Scatterplot Matrices

Our primary example in this chapter is a troublesome regression using World Bank (1987) data on 109 countries (*nations.dta*):

```
. describe

Contains data from c:\stustata\nations
  Obs:      109 (max=  2620)              Data on 109 countries
  Vars:      15 (max=    99)
  Width:     33 (max=   200)
   1. country      str8    %9s            Country
   2. pop          float   %9.0g          1985 Population, Millions
   3. gnpgro       float   %9.0g          Annual GNP Growth % 65-85
   4. gnpcap       int     %8.0g          Per Capita GNP 1985
   5. chldmort     byte    %8.0g          Child Mortality (1-4year) 85
   6. birth        byte    %8.0g          Crude Birth Rate/1000 Populatio
   7. life         byte    %8.0g          Life Expectancy @ Birth 85
   8. energy       int     %8.0g          Energy Consumpt/cap kg oil equi
   9. death        byte    %8.0g          Crude Death Rate/1000 Populatio
  10. infmort      int     %8.0g          Infant Mortality (<1year) 85
  11. food         int     %8.0g          Daily Calories/cap 1985
  12. school1      int     %8.0g          Total Primary Enrol. % Age-Grou
  13. school2      byte    %8.0g          Total Secondary Enrol % Age-Gro
  14. school3      byte    %8.0g          Higher Education Enrol % Age-Gr
  15. urban        byte    %8.0g          % Population Urban 1985
Sorted by:
```

We will examine how the crude birth rate relates to per capita gross national product and the child mortality rate. A correlation matrix indicates moderate-strength linear relations among these variables:

```
. correlate gnpcap chldmort birth
(obs=109)
         |   gnpcap chldmort    birth
---------+---------------------------
  gnpcap |   1.0000
chldmort |  -0.5047   1.0000
   birth |  -0.6263   0.7773   1.0000
```

The corresponding scatterplot matrix, however, tells a more complicated story (Figure 8.1):

```
. graph gnpcap chldmort birth, matrix half label
```

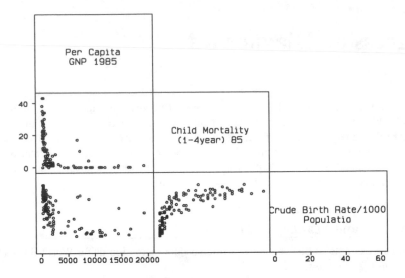

**Figure 8.1**

All three scatterplots in Figure 8.1 reveal distinct curvilinearity; outliers and heteroscedasticity (nonconstant variance) may also present problems.

Thoughtful analysts would stop at this point, and proceed no further with linear model-fitting until they found a way to accommodate the curvilinearity visible in Figure 8.1. Chapter 9 shows how power transformations might be used. For purposes of illustrating diagnostic techniques, however, this chapter will proceed with obviously misguided linear model-fitting, regressing birth rate (*birth*) on per capita GNP (*gnpcap*) and child mortality rate (*chldmort*). We could perform this regression by typing:

. **regress** *birth gnpcap chldmort*

Alternatively, the **fit** command accomplishes the exact same regression:

. **fit** *birth gnpcap chldmort*

| Source | SS | df | MS |  | |
|--------|-----|-----|-----|---|---|
| | | | | Number of obs = | 109 |
| | | | | F( 2, 106) = | 111.41 |
| Model | 13604.1783 | 2 | 6802.08913 | Prob > F   = | 0.0000 |
| Residual | 6471.96852 | 106 | 61.0563068 | R-square   = | 0.6776 |
| | | | | Adj R-square = | 0.6715 |
| Total | 20076.1468 | 108 | 185.890248 | Root MSE   = | 7.8139 |

| birth | Coef. | Std. Err. | t | P>\|t\| | [95% Conf. Interval] | |
|-------|-------|-----------|---|---------|---------------------|---|
| gnpcap | -.0009631 | .000196 | -4.914 | 0.000 | -.0013517 | -.0005745 |
| chldmort | .7512455 | .0775466 | 9.688 | 0.000 | .5975018 | .9049893 |
| _cons | 28.37934 | 1.42717 | 19.885 | 0.000 | 25.54983 | 31.20884 |

While performing the regression, **fit** unobtrusively stores some calculations that will make subsequent diagnostic work easier. The output indicates that *gnpcap* and *chldmort* both have statistically significant effects, and together explain about 67% of the variance in birth rates.

# Residual Versus Predicted *Y* Plots

Chapter 6 introduced residual versus predicted *Y* plots.  After **fit**, we can draw such plots (also called "residual versus fitted" plots) automatically, using the **rvfplot** command (Figure 8.2):

```
. rvfplot, yline(0) twoway box ylabel xlabel
```

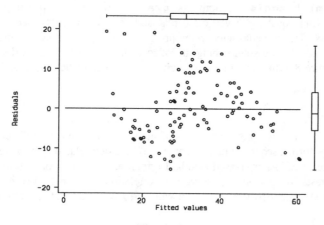

**Figure 8.2**

Residuals from the birth rate regression (Figure 8.2) show further evidence of the curvilinearity visible in Figure 8.1.  The residuals are mostly negative (too-high predictions) at low predicted (fitted) birth rates, positive at middle predicted values, and turn negative again at high values.  Several outliers appear at top left in Figure 8.2.

Figure 8.2 does not show much evidence of heteroscedasticity, another problem common in aggregate data.  To be sure, we might graph absolute residuals against predicted *Y* (Figure 8.3):

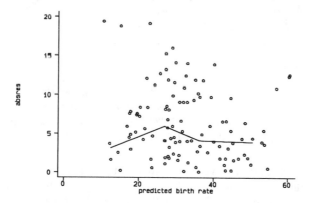

**Figure 8.3**

```
. predict yhat
. label variable yhat "Predicted birth rate"
. predict resid, resid
. label variable resid "Residual"
. generate absres = abs(resid)
. graph absres yhat, connect(m) bands(4) ylabel xlabel
```

The `connect(m) bands(4)` options ask `graph` to divide the range of $X$-variable (here, *yhat*) values into four equal-width vertical bands, and connect the cross-medians of each band with line segments. Heteroscedasticity, when present, should produce systematic change in this band regression line. (Lowess smoothing, introduced in Chapter 9, also works for this sort of exploration.) Figure 8.3, like Figure 8.2, contains no evidence of heteroscedasticity.

## Autocorrelation

Autocorrelation, or correlation between values of the same variable across different cases, often occurs when cases form a sequence in time (or space). Autocorrelated regression errors produce inefficient parameter estimates, biased standard errors, and invalid $t$ or $F$ tests. The Durbin-Watson statistic tests for first-order positive autocorrelation of regression residuals. To obtain a Durbin-Watson statistic, perform regression with `regdw` instead of `regress` or `fit`:

```
. regdw birth gnpcap chldmort
```

| Source | SS | df | MS | | Number of obs = | 109 |
|---|---|---|---|---|---|---|
| | | | | | F( 2, 106) = | 111.41 |
| Model | 13604.1783 | 2 | 6802.08913 | | Prob > F    = | 0.0000 |
| Residual | 6471.96852 | 106 | 61.0563068 | | R-square    = | 0.6776 |
| | | | | | Adj R-square = | 0.6715 |
| Total | 20076.1468 | 108 | 185.890248 | | Root MSE    = | 7.8139 |

| birth | Coef. | Std. Err. | t | P>|t| | [95% Conf. Interval] | |
|---|---|---|---|---|---|---|
| gnpcap | -.0009631 | .000196 | -4.914 | 0.000 | -.0013517 | -.0005745 |
| chldmort | .7512455 | .0775466 | 9.688 | 0.000 | .5975018 | .9049893 |
| _cons | 28.37934 | 1.42717 | 19.885 | 0.000 | 25.54983 | 31.20884 |

```
Durbin-Watson Statistic =   1.834577
```

Given $n = 109$ (approximately 100) and two $X$ variables, the .05-level critical values for Durbin-Watson's $d$ statistic (looked up in tables at the back of most regression texts) are $d_L = 1.63$ and $d_U = 1.72$. `regdw` reports $d = 1.834577$, above $d_U$, so we fail to reject the null hypothesis of no autocorrelation at the $\alpha = .05$ level. That is, we find no significant autocorrelation among the residuals.

This conclusion misleads, however. A Durbin-Watson test refers to the data as presently sorted, and these nations are in alphabetical order. Their sequence begins Algeria, Argentina, Australia, Austria, Bangladesh .... Of course there is no autocorrelation between residuals of adjacent countries in this series, because they have no "real" adjacency. On the other hand, if the countries were sorted geographically, we might see substantial autocorrelation. This example makes two points about autocorrelation tests:

1.    they reach different results depending on how the cases are sorted; and
2.    unless cases form a clear sequence, as with time series, the sorting order is arbitrary.

For an unambiguous example of autocorrelation, we turn to time series data on water use for Concord, New Hampshire, over months from 1970–1981 (*concord2.dta*). During 1979–81, Concord faced a drought and water shortage. Local government responded with a conservation campaign, educating citizens about the problem and urging them to conserve water.

```
Contains data from c:\stustata\concord2.dta
  Obs:   137 (max=  2620)               Concord Water Department (1982)
  Vars:    7 (max=    99)
 Width:   17 (max=   200)
  1. year          byte   %8.0g              Year
  2. month         byte   %8.0g    monthlbl  Month
  3. time          int    %8.0g              index number 1-137
  4. water         float  %9.0g              average daily water use
  5. temp          float  %9.0g              average monthly temperature
  6. rain          float  %9.0g              precipitation in inches
  7. educ          byte   %8.0g              conservation campaign dummy
Sorted by:  time
```

`. list year month water temp rain educ in 1/6`

```
      year    month     water      temp      rain     educ
 1.     70      Jan   3.807097        11        .4        0
 2.     70      Feb   4.068929      22.6      4.27        0
 3.     70      Mar   4.225484      30.7      2.78        0
 4.     70      Apr      4.085      45.8      3.38        0
 5.     70      May   3.830323      57.3      3.04        0
 6.     70      Jun      4.517      63.1      2.26        0
```

We regress average daily water use on temperature, precipitation, and a dummy variable representing the conservation education campaign:

`. regdw water temp rain educ`

```
  Source |       SS       df       MS              Number of obs =     137
---------+------------------------------           F(  3,    133) =   21.57
   Model | 8.40906094      3  2.80302031           Prob > F      = 0.0000
Residual | 17.2802366    133  .129926591           R-square      = 0.3273
---------+------------------------------           Adj R-square  = 0.3122
   Total | 25.6892976    136  .188891894           Root MSE      = .36045

-----------------------------------------------------------------------------
   water |     Coef.   Std. Err.       t     P>|t|     [95% Conf. Interval]
---------+-------------------------------------------------------------------
    temp |   .0128571   .0016975      7.574   0.000     .0094995    .0162147
    rain |  -.0474281    .021229     -2.234   0.027    -.0894182    -.005438
    educ |  -.2469767   .1134846     -2.176   0.031    -.4714448   -.0225086
   _cons |   3.828001   .1006446     38.035   0.000      3.62893    4.027072
-----------------------------------------------------------------------------
Durbin-Watson Statistic =   .5348529
```

Given $n = 137$ (approximately 100) and three $X$ variables, the .05-level critical values are $d_L = 1.61$ and $d_U = 1.74$. **regdw** reports $d = .5348529$, well below $d_L$, so we can reject the null hypothesis of no autocorrelation at the $\alpha = .05$ level. That is, we find significant autocorrelation among the residuals, and so cannot trust the output's standard errors, $t$ tests, or $F$ test.

Graphs of residuals versus time provide another way to look for autocorrelation. In these data the variable *time* measures months since December 1969 (Figure 8.4).

```
. predict e, resid
. label variable e "residual"
. graph e time, connect(l) symbol(i) yline(0) ylabel xlabel
       b2(months since December 1969)
```

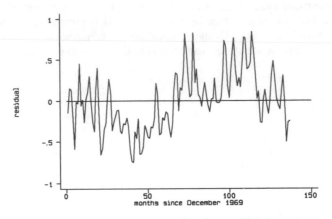

**Figure 8.4**

Sometimes smoothing allows a better view of underlying patterns, by clearing away random noise. **egen** can produce moving averages (also called running means) of any span. For example, to smooth the water-use residuals (*e*) by moving averages of span 5 (Figure 8.5):

```
. egen smooth = ma(e), nomiss t(5)
. label variable smooth "smoothed residual"
. graph smooth time, connect(1) symbol(i) yline(0) ylabel xlabel yscale(-1,1)
      b2(months since December 1969)
```

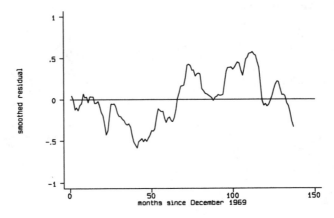

**Figure 8.5**

Figures 8.4 and 8.5 reveal the pattern of autocorrelation in these residuals: our regression model's predictions tend to be too high (causing negative residuals) in the first half of the data and too low (positive residuals) in much of the rest. At the end, during the conservation campaign, residuals again tend to be negative. Such graphs can add a qualitative, descriptive element to the Durbin-Watson finding of significant autocorrelation.

# Nonnormality

The usual regression $t$ and $F$ tests assume normally distributed errors. Furthermore, ordinary least squares (OLS) is more efficient than other unbiased estimators if errors are normal. Sample residuals provide evidence about the plausibility of assuming normal errors.

Chapters 3 and 4 described some graphical and statistical normality checks that can be applied to residuals as they can to any other variable. For example, if our model residuals are called *resid*, we might try any of the following diagnostic analyses:

. **graph** *resid*, **norm**      Histogram with superimposed normal curve. Options control the number of bins (bars) in the histogram, and the mean and standard deviation for the normal curve. To produce a histogram with 12 bars, and a normal curve with mean 0 and standard deviation 5:
  **graph** *resid*, **bin(12) norm(0,5)**

. **graph** *resid*, **box**      Boxplots highlight some signs of nonnormality: skew and outliers. The absence of skew or outliers does not indicate normality, however—perfectly symmetrical distributions could still be nonnormal. To identify individual outliers in a boxplot, use the **symbol([varname])** option, where *varname* is a variable holding case ID numbers or names. For example:
  **graph resid, box symbol([country])**

. **qnorm** *resid*      Quantile-normal plots show any departures from normality, with much finer detail than histograms or boxplots.

. **summarize** *resid*, **detail**  For a simple check of symmetry, compare median (50th percentile) with mean. Skewness measures the extent and direction of asymmetry: skewness > 1 indicates positive skew. Kurtosis measures tail weight: kurtosis > 3 indicates heavier-than-normal tails.

. **lv** *resid*      Letter-values displays provide a wealth of information. Drift in midsummaries indicates progressively changing skew.

. **sktest** *resid*      This command and its relatives, **sktestd**, **sfrancia**, and **swilk**, provide formal tests of the null hypothesis that the sample came from a normal population. See *Stata Technical Bulletin* 5 for a Monte Carlo evaluation of these four normality tests.

Unfortunately, nonnormality tends to be most consequential in small samples, where normality tests lack power.

If residual analysis casts any doubt on normality assumptions, we should consider different model-fitting strategies. Unlike the classical parametric methods, nonparametric methods (Chapter 4) and robust methods (Chapter 10) do not assume normality. Alternatively, a simple power transformation (Chapters 2, 9) may bring skewed data closer to normality.

# *DFBETAS*

*DFBETAS* are case statistics measuring how much a regression coefficient would change if that particular case were deleted. Used after **fit**ting a regression model, the **dfbeta** command generates *DFBETAS* for the listed $X$ variables. For illustration, we return to the crossnational data in *nations.dta*:

```
. use c:\stustata\nations, clear
. quietly fit birth gnpcap chldmort
. dfbeta gnpcap chldmort
DFgnpcap:   DFbeta(gnpcap)
DFchldmo:   DFbeta(chldmort)
```

Preceding any Stata command by the word **quietly** tells Stata to execute that command, but not display any output onscreen. The **dfbeta** command created two new variables, automatically named *DFgnpcap* and *DFchldmo*.

Influential cases have large *DFBETAS*. **summarize** lists low and high extremes:

```
. summarize DFgnpcap DFchldmo
```

| Variable | Obs | Mean | Std. Dev. | Min | Max |
|---|---|---|---|---|---|
| DFgnpcap | 109 | .0016865 | .1432527 | -.1663877 | 1.072631 |
| DFchldmo | 109 | -.0004444 | .1097267 | -.5188906 | .3429605 |

Since the largest *DFgnpcap* equals 1.072631, we know that at least one case is so influential that it pulls the coefficient on *gnpcap* up by about 1.07 standard errors. To list countries that shift the coefficient on *gnpcap* by half a standard error or more (that is, have absolute values of *DFgnpcap* above .5), type:

```
. list country DFgnpcap if abs(DFgnpcap) > .5
```

| | country | DFgnpcap |
|---|---|---|
| 89. | Kuwait | .6340666 |
| 90. | UnArEmir | 1.072631 |

The oil exports and small populations of Kuwait and the United Arab Emirates give them very high per capita GNP. Most high-GNP countries have low birth rates, but these two do not; their unusual combination of high GNP with high birth rate makes them influential.

Similarly, we could list countries that influence the coefficient on child mortality by more than half a standard error:

```
. list country DFchldmo if abs(DFchldmo) > .5
```

| | country | DFchldmo |
|---|---|---|
| 4. | Mali | -.5188906 |
| 26. | SierraLe | -.5157053 |

Mali and Sierra Leone both pull this coefficient down by about half a standard error.

To see the effects of these influential cases, we could redo the regression without them:

```
. fit birth gnpcap chldmort if abs(DFgnpcap) < .5 & abs(DFchldmo) < .5
```

| Source | SS | df | MS | | Number of obs = | 105 |
|---|---|---|---|---|---|---|
| | | | | | F( 2, 102) = | 141.04 |
| Model | 14392.1647 | 2 | 7196.08236 | | Prob > F = | 0.0000 |
| Residual | 5204.06385 | 102 | 51.0202338 | | R-square = | 0.7344 |
| | | | | | Adj R-square = | 0.7292 |
| Total | 19596.2286 | 104 | 188.425275 | | Root MSE = | 7.1428 |

| birth | Coef. | Std. Err. | t | P>|t| | [95% Conf. Interval] | |
|---|---|---|---|---|---|---|
| gnpcap | -.001253 | .0002013 | -6.224 | 0.000 | -.0016523 | -.0008537 |
| chldmort | .7951894 | .0788009 | 10.091 | 0.000 | .6388882 | .9514906 |
| _cons | 28.70069 | 1.370117 | 20.948 | 0.000 | 25.98307 | 31.41831 |

Without these four influential cases, both regression coefficients are steeper, and $R^2$ increases from .6776 to .7344. We see stronger effects because the influential cases were exceptions to the general patterns of the data. That does not necessarily mean the influential cases *ought* to be dropped, however—such decisions should be made on substantive grounds.

The univariate outlier-detection methods of Chapters 3–4 can help in identifying unusually influential cases. For example, a letter-values display:

```
. lv DFgnpcap
```

```
 #      109                    DFbeta(gnpcap)
                 ---------------------------------
 M      55   |              -.0099486          |      spread    pseudosigma
 F      28   | -.0711336   -.0291451   .0128434 |      .083977     .0631476
 E      14.5 | -.0936585   -.0137412   .0661761 |     .1598345     .0708228
 D       7.5 | -.1228482    .0145204   .151889  |     .2747372     .0909854
 C       4   |  -.150601    .0307837   .2121685 |     .3627695     .0990533
 B       2.5 | -.1562595    .1524896   .4612387 |     .6174982     .1500573
 A       1.5 | -.1631357    .3451064   .8533486 |    1.016484      .2207936
         1   | -.1663877    .4531214  1.072631  |    1.239018      .2486109
             |                                  |
             |                                  |   # below     # above
inner fence  | -.1970991               .1388088 |      0           8
outer fence  | -.3230645               .2647743 |      0           3
```

Among *DFBETAS* on *gnpcap*, we find 8 outliers, 3 of them severe. These 8 countries pull the coefficient on *gnpcap* upward, closer to zero. To identify individual outliers, we can **list** them (using inner fence as a cutoff) or draw a boxplot (Figure 8.6):

```
. list country DFgnpcap if DFgnpcap > .1388

           country   DFgnpcap
102.         U_S_A   .1511005
103.          Oman   .1526775
104.       Hungary   .153791
105.         China   .1718739
106.         Libya   .2121685
107.      SauArabi   .2884108
108.        Kuwait   .6340666
109.      UnArEmir  1.072631
```

```
. graph DFgnpcap, box symbol([country]) psize(140) ylabel
```

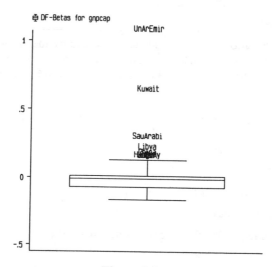

**Figure 8.6**

We might also employ oneway scatterplots or quantile-normal plots (Chapter 3) to look for gaps in the distribution of *DFgnpcap*, separating a few unusually influential cases from the rest.

    **avplots** (Professional Stata only) produces a set of partial regression leverage plots, also called added-variable plots, for all *X* variables in the previous **fit**. For example (Figure 8.7):

```
. quietly fit birth gnpcap chldmort
. avplots
```

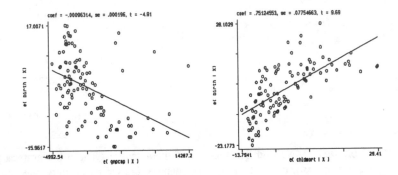

**Figure 8.7**

Leverage-plot slopes equal the corresponding partial regression coefficients on *gnpcap* (left) and *chldmort*. Kuwait and the United Arab Emirates are at upper right in the left-hand plot. We can see why these two pull the partial regression line up, making it less steep. For individual leverage plots (Student or Professional Stata) type **avplot gnpcap** or **avplot chldmort** instead of just **avplots**.

## Other Case Statistics

After **regress**, **fit**, **anova**, or other estimation procedures, the **predict** command can generate case statistics such as predicted values, several kinds of residuals or standard errors, leverage, and Cook's *D*. See page 75 for a complete list of **predict** options.

```
. predict newvar, hat
```

creates a new variable called *newvar* (or whatever name we give it), containing values of hat matrix diagonals or leverage. Leverage reflects potential influence, resulting from an unusual combination of *X*-variable values.

```
. predict newvar, stdr
```

estimates the standard deviation of the *i*th residual.

```
. predict newvar, rstandard
```

calculates standardized residuals: the *i*th-case residual divided by its standard error.

```
. predict newvar, rstudent
```

Studentized residuals are equivalent to a *t* test for whether the *i*th case significantly shifts the regression intercept, and hence should be considered an outlier.

```
. predict newvar, cooksd
```

Cook's $D$ measures how much the $i$th case influences the regression equation as a whole. (In contrast, *DFBETAS* measures how much the $i$th case influences a specific regression coefficient.)

For example, after the birth rate regression we might obtain these case statistics:

```
. predict yhat
. predict resid, resid
. predict h, hat
. predict student, rstudent
. predict D, cooksd
```

**summarize** will show us the extreme values:

```
. summ resid h student D
```

| Variable | Obs | Mean | Std. Dev. | Min | Max |
|---|---|---|---|---|---|
| resid | 109 | 2.73e-09 | 7.741165 | -15.25247 | 19.42905 |
| h | 109 | .0275229 | .0220845 | .0101312 | .1435016 |
| student | 109 | .0014921 | 1.018261 | -2.000394 | 2.770001 |
| D | 109 | .0131553 | .0432924 | 2.21e-07 | .4031392 |

Leverage ($h$) theoretically can range from $1/n$ to 1. The largest value in this sample, $h = .1435$ (United Arab Emirates) indicates only a moderate potential for influence.

Special tables could be used to determine whether the case with the largest absolute studentized residual (*student*) constitutes a significant outlier. Alternatively, we could apply the Bonferroni inequality: $\max|t|$ is significant at level $\alpha$ if $t$ is significant at $\alpha/n$. In this example, we have $\max|t| = 2.77$ (U.A.E. again) and $n = 109$. For U.A.E. to be a significant outlier (cause a significant shift in intercept) at $\alpha = .05$, $t = 2.77$ must be significant at $.05/109$:

```
. display .05/109
.00045872
```

Stata's **tprob()** function can approximate the probability of $|t| > 2.77$, given $df = n-K-1 = 109-3-1 = 105$:

```
. display tprob(105,2.77)
.00663015
```

Since the obtained $P$-value ($P = .00663015$) is not below $\alpha/n = .00045872$, U.A.E. is not a significant outlier at $\alpha = .05$.

Studentized residuals measure the $i$th case's influence on the $Y$-intercept, and *DFBETAS* measure the $i$th case's influence on the $k$th regression coefficient. Cook's $D$ provides an overall measure of influence, reflecting the $i$th case's impact on all coefficients in the model (or, equivalently, on all $n$ predicted $Y$ values). To list the 4 most influential cases, for instance:

```
. sort D
. list country D in -4/1
```

| | country | D |
|---|---|---|
| 106. | SierraLe | .1004216 |
| 107. | Mali | .1024002 |
| 108. | Kuwait | .1611289 |
| 109. | UnArEmir | .4031392 |

The **in -4/1** qualifier tells Stata to list only the fourth-from-last ($-4$) through last (lowercase letter "l") observations.

Graphical displays for Cook's $D$ might include boxplots to spot the influence outliers:

```
. graph D, box
```

or a residual versus predicted $Y$ plot, with symbols proportional to $D$ (not shown):

```
. graph resid yhat [iweight=D]
```

A nicer version of the latter graph results if we rescale $D$ values before using them as weights (Figure 8.8):

```
. generate Dstar = (99/4)*D*(D+1)^2+1
. replace Dstar = 100 if D > 1
. label variable Dstar "Cook's D rescaled"
. graph resid yhat [iweight=Dstar], yline(0) ylabel xlabel
```

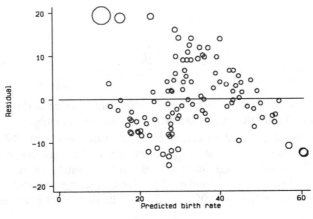

**Figure 8.8**

The rescaling ensures that cases with $D \geq 1$ will plot with about 100 times the area of the least-influential cases. In Figure 8.8, United Arab Emirates appears at upper left, much more influential than any other country.

Several kinds of diagnostic graphs become available after **fit**. Confusingly, the statistical literature calls many of these graphs by more than one name. Stata's post-**fit** graphing commands are:

|  |  |
|---|---|
| **avplot** | individual leverage or added-variable plot |
| **avplots** | all leverage or added-variable plots (Professional Stata only) |
| **cprplot** | partial residual or component-plus-residual plot |
| **acprplot** | Mallows' augmentation of component-plus-residual plot |
| **lvr2plot** | L-R or leverage vs. squared residual plot |
| **rvfplot** | $e$ vs. *Yhat* or residual vs. fitted plot |
| **rvpplot** | $e$ vs. $X$, independent variable, carrier, or residual vs. predictor plot |

Type **help fit** for the syntax of these commands.

An L-R plot, for example, graphically shows which cases combine potential influence with poor fit to the model (Figure 8.9):

```
. quietly fit birth gnpcap chldmort
. lvr2plot, border
```

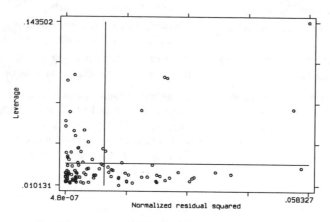

**Figure 8.9**

Points above the horizontal line have greater-than-average leverage; points right of the vertical line have larger-than-average residuals. Thus points at upper left have high leverage but small residuals: either they agree with the main pattern of the data, or they have enough influence to control the regression. Conversely, the leverage points at upper right fit the regression predictions less closely. To identify the outliers, try `lvr2plot, border symbol([country])` (results not shown here).

    `predict` works the same way after either `fit` or `regress`. After `fit`, though, we could instead use the similar `fpredict` command. `fpredict` does many of the same things as `predict` (except out-of-sample predictions), but it also permits several new options. A complete list of `fpredict` options follows. The list notes suggested "size-adjusted cutoffs" for influence statistics, meant to flag unusually influential cases:

. `fpredict` *newvar*                predicted $Y$; `fpredict` does <u>not</u> predict $Y$ values for cases with missing $Y$ (unlike `predict`)

. `fpredict` *newvar*, `cooksd`        Cook's $D$ influence measure; look closely at cases with $D > 4/n$

. `fpredict` *newvar*, `covratio`    *COVRATIO* influence measure; look closely if $|COVRATIO - 1| \geq 3K/n$

. `fpredict` *newvar*, `dfbeta(xvar)`   *DFBETAS* for specified *xvar*; look closely if $DFBETAS > 2/\sqrt{n}$

. `fpredict` *newvar*, `dfits`        *DFFITS*; look closely at cases if $DFFITS > 2\sqrt{K/n}$

. `fpredict` *newvar*, `hat`          leverage or hat matrix diagonals; look closely if $h > 2K/n$

. `fpredict` *newvar*, `residuals`    residuals

. `fpredict` *newvar*, `rstandard`   standardized residuals

. `fpredict` *newvar*, `rstudent`     studentized residuals (see page 105 regarding interpretation)

. `fpredict` *newvar*, `stdf`         standard error of forecast, for individual-case predicted values

. `fpredict` *newvar*, `stdp`         standard error of prediction, for conditional-mean predictions

```
. fpredict newvar, stdr
```
standard error of residuals

```
. fpredict newvar, welsch
```
Welsch distance influence measure; look closely at cases with *Welsch* > 3/*K*

In using `fpredict`, substitute a new variable name of your choosing for the "*newvar*" shown above. Typing `help fit` obtains the full syntax for `fpredict` and its options.

Although these analyses found several disproportionately influential cases, their influence is not great in absolute terms. The curvilinear pattern clearly visible in graphs like Figures 8.7–8.8 presents a more serious challenge to the validity of this regression. Fitting a linear model to obviously curved data is misguided, and we should address this problem before worrying about influence. Chapter 9 illustrates an approach that sometimes reduces both problems at once.

## Multicollinearity

If perfect multicollinearity (linear relation) exists among predictor variables, regression equations become unsolvable. Stata handles this by warning the user, then automatically dropping one of the offending variables. High (but not perfect) multicollinearity also causes problems, though its symptoms are less obvious. When we add to the analysis a new $X$ variable that is strongly related to $X$ variables already in the model, symptoms include:

1.   substantially higher standard errors, and correspondingly lower $t$ statistics;

2.   unexpected coefficient magnitudes, signs, or sign reversals; and

3.   nonsignificant coefficients despite a high $R^2$.

The matrix of correlations between estimated regression coefficients provides one diagnostic for multicollinearity. This matrix can be displayed after **regress**, **fit**, **anova**, or other estimation procedures by typing:

```
. correlate, _coef
         |   gnpcap chldmort      _cons
---------+------------------------------
  gnpcap|   1.0000
chldmort|   0.5047   1.0000
   _cons|  -0.7117  -0.7627   1.0000
```

High correlations between any pair of coefficients indicates the presence of multicollinearity. In this example, correlations among coefficients appear to be moderate, suggesting that multicollinearity is not a great problem.

For a more detailed look, regress each $X$ variable on all of the others and check the resulting $R^2$:

```
. quietly regress chldmort gnpcap
. display _result(7)
.25474522
```

After a regression, `_result(7)` holds the $R^2$ (consult *Stata Reference Manual* for a complete list of `_result` functions). The last two commands thus illustrate how to obtain $R^2$ alone, without further unwanted output. We see that *gnpcap* explains only about 25% of the variance of *chldmort*. Since 75% of *chldmort*'s variance is independent of *gnpcap*, multicollinearity should be no problem here.

Similarly, to see the $R^2$ from regressing *gnpcap* on *chldmort*, type:

```
. quietly regress gnpcap chldmort
. display _result(7)
.25474522
```

This example involves only two $X$ variables, so the regression of $X_1$ on $X_2$ yields the same $R^2$ as the regression of $X_2$ on $X_1$. We would see more variety with three or more $X$ variables.

## Also Type help

| | |
|---|---|
| corc | regression with Cochrane-Orcutt correction for serial correlation |
| correlate | correlation or covariance matrix of variables or coefficients |
| diagplot | distribution diagnostic plots |
| estimate | hold results from last model fitting |
| fit | OLS regression with diagnostics |
| graph | general graphing command |
| predict | predicted values, residuals, or diagnostic statistics |
| regress | OLS regression; also WLS, instrumental variables, 2SLS |
| test | $F$ tests of user-specified hypotheses |

# 9
# Fitting Curves

This chapter describes three broad approaches to fitting curves:
1. nonparametric regression;
2. linear regression with transformed variables ("curvilinear regression"); and
3. nonlinear regression.

Nonparametric regression often serves as an exploratory technique, without an explicit model. Curvilinear regression provides a simple way to describe curvilinear relations with intrinsically linear models. Nonlinear regression, on the other hand, refers to a different class of techniques needed for intrinsically nonlinear models.

## Nonparametric Methods: Band Regression and Lowess Smoothing

Nonparametric regression methods generally do not yield an explicit regression equation. They are primarily graphical tools for displaying the relation, possibly nonlinear, between $Y$ and $X$. Stata can perform a simple nonparametric regression, called band regression, as part of any two-variable scatterplot. For illustration, consider these sobering Cold War data (*missile.dta*) from MacKenzie (1990). The cases are 48 types of long-range nuclear missiles, deployed by the U.S. and Soviet Union during their arms race, 1958–1990:

```
Contains data from c:\stustata\missile.dta
  Obs:    48  (max=  2620)
  Vars:    6  (max=    99)            MacKenzie (1990) missile data
 Width:   25  (max=   200)
   1. missile    str15   %15s                 Missile
   2. country    byte    %8.0g     soviet     US or Soviet missile?
   3. year       int     %8.0g                Year of first deployment
   4. type       byte    %8.0g     type       ICBM or submarine-launched?
   5. range      int     %8.0g                Range in nautical miles
   6. CEP        float   %9.0g                Circular Error Probable (miles)
Sorted by:  country  year
```

`. list in 1/10`

```
            missile   country    year     type     range      CEP
  1.        Atlas D    U.S.      1959     ICBM        .        1.8
  2.      Polaris A1   U.S.      1960     SLBM      1200         2
  3.        Atlas E    U.S.      1961     ICBM        .          1
  4.        Atlas F    U.S.      1961     ICBM        .          1
  5.     Minuteman 1   U.S.      1962     ICBM        .        1.1
  6.      Polaris A2   U.S.      1962     SLBM      1500         2
  7.        Titan 1    U.S.      1962     ICBM        .        .65
  8.        Titan 2    U.S.      1963     ICBM        .        .65
  9.      Polaris A3   U.S.      1964     SLBM      2500         .5
 10.     Minuteman 2   U.S.      1966     ICBM        .        .26
```

Variables in *missile.dta* include an accuracy measure called "Circular Error Probable" (*CEP*). *CEP* represents the radius of a bullseye within which 50% of the missile's warheads should land. Year by year, scientists on both sides worked to improve accuracy (Figure 9.1):

`. graph CEP year, bands(8) connect(m) ylabel xlabel`

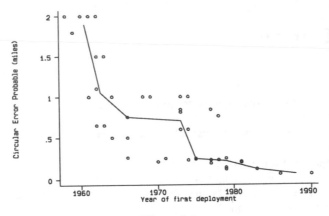

**Figure 9.1**

Figure 9.1 shows *CEP* declining (accuracy increasing) over time. The options **bands(8)** and **connect(m)** tell **graph** to divide the scatterplot into 8 vertical bands, and draw line segments connecting cross-medians within each band. This band regression traces how the median of *CEP* changes with *year*.

Nonparametric regression does not ask us to specify a relationship's functional form (linear or otherwise) in advance. Instead, it explores the data with an "open mind." This often uncovers interesting results, as when we view U.S. and Soviet missile accuracy separately (Figure 9.2):

```
. graph CEP year, connect(m) bands(8) ylabel xlabel by(country)
```

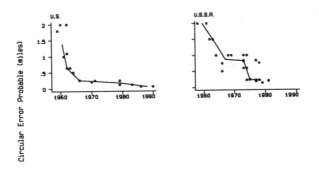

**Figure 9.2**

The shapes of the two curves in Figure 9.2 differ substantially. U.S. missiles became dramatically more accurate in the 1960s, permitting a shift to smaller warheads. Three or more small warheads would fit on the same size missile that formerly carried one large warhead. The accuracy of Soviet missiles improved more slowly, apparently stalling during the late 60s–early 70s, and remained a decade or so behind their American counterparts. To make up for this accuracy disadvantage, Soviet strategy emphasized large rockets with huge warheads.

A more elaborate nonparametric regression technique called lowess smoothing has recently gained popularity. Lowess stands for **locally weighted scatterplot smoothing**. For a lowess-smoothed graph of *CEP* against *year*, for U.S. missiles only, type:

. **ksm** *CEP year* **if** *country==0*, **lowess**

**ksm** understands all the usual Stata **graph** options, and with a **gen()** option, it also creates smoothed predicted values as a new variable (Figure 9.3):

. **ksm** *CEP year* **if** *country==0*, **lowess bwidth(.4) ylabel xlabel symbol(Oi)**
      **gen(***lsCEP***)**

. **label variable** *lsCEP* **"Lowess Smoothed CEP"**

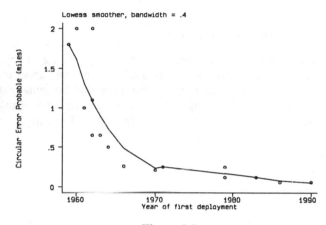

**Figure 9.3**

Like Figure 9.2, Figure 9.3 shows U.S. missile accuracy improving rapidly during the 1960s, and progressing at a more gradual rate in the 1970s and 1980s. The **bwidth(.4)** option specifies the lowess bandwidth: the fraction of the data used in smoothing each point. Default is **bwidth(.8)**. The closer bwidth is to 1, the greater the degree of smoothing.

Lowess predicted (smoothed) *Y* values for *n* cases result from *n* weighted regressions. Within the *i*th band (the *i*th set of bwidth $\times$ *n* observations, centered on case *i* ), the *j*th case receives weight ($w_j$) according to a tricube function:

$$w_j = (1 - |u_j|^3)^3 \qquad \text{if } |u_j| < 1$$
$$w_j = 0 \qquad\qquad\quad \text{if } |u_j| \geq 1$$

where

$$u_j = (X_i - X_j)/d_i$$

$$d_i = \text{bwidth} \times n/2, \text{ rounded to nearest integer}$$

Smaller, uncentered bands are used in calculating weights towards the end points. Weights equal 1 at the center of each band, but fall to 0 at its boundaries. Chambers et al. (1983) and Cleveland (1985) provide further details and examples.

**ksm** is a flexible command, described in more detail by Royston (1991). In addition to the usual **graph** options, **ksm** offers these special options:

| | |
|---|---|
| **line** | for running-line least squares smoothing; default is running mean |
| **weight** | for Cleveland's tricube weighting function; default is unweighted |

| | |
|---|---|
| lowess | equivalent to specifying both options `line weight` |
| bwidth(#) | specifies the bandwidth. Centered subsets of `bwidth` $\times n$ observations are used for smoothing; default is `bwidth(.8)` |
| logit | transforms smoothed values to logits |
| adjust | adjusts the mean of smoothed values to equal the mean of the original $Y$ variable; like `logit`, useful with dichotomous $Y$ |
| nograph | suppresses display of the graph; often used with `gen()` |
| gen(newvar) | creates *newvar* containing smoothed values of $Y$ |

To fit $n$ data points, lowess smoothing requires $n$ weighted regressions, and so proceeds slowly with large samples or old computers.

## Regression with Transformed Variables

Chapter 6 introduced regression with transformed variables—regressing $Y$ on log $X$, or on $X$ and $X^2$, and so forth. (See Chapter 2, and also `help ladder`, for more about transformations.) Transformations extend the scope of linear models, but many students initially object that they seem like hocus-pocus or unnatural fiddling with the data. To help convince skeptics, here is another simple example where transformation obviously works. The data (*tornado.dta*) come from the Council on Environmental Quality (1988):

```
Contains data from c:\stustata\tornado.dta
  Obs:    71 (max=  2620)            Environmental Quality 1987-88
 Vars:     5 (max=    99)
Width:    11 (max=   200)
  1. year         int      %8.0g       Year
  2. tornado      int      %8.0g       Number of tornados
  3. lives        int      %8.0g       Number of lives lost
  4. hurric       byte     %8.0g       Number N. Atlantic hurricanes
  5. avlost       float    %9.0g       Average lives lost/tornado
Sorted by:
```

The average number of lives lost per tornado declined over 1916–1986, due to more effective warnings and our ability to detect more tornados now, even ones that cause little damage. A linear regression line does not well describe this trend (Figure 9.4). First, the actual trend appears curvilinear, leveling off at small loss of life after the mid-1950s, while the regression line predicts negative deaths. Second, average tornado deaths exhibit much more variation in the early years than later—clear heteroscedasticity. (Try a Durbin-Watson test, `regdw`, to see whether serial correlation also troubles this regression.)

The relation becomes linear, and heteroscedasticity vanishes, if we work instead with the log of average lives lost (Figure 9.5):

```
. generate lavlost = ln(avlost)
. label variable lavlost "Log of average lives lost"
. quietly regress lavlost year
. predict yhat
. graph lavlost yhat year, connect(.s) symbol(Oi) ylabel
        xlabel(1920,1930,1940,1950,1960,1970,1980)
```

Since we regressed log lives lost on *year*, the predicted values (*yhat*) are also measured in logarithmic units. Return these predicted values to their natural units (lives lost) by inverse transformation, in this case exponentiating (*e* to power) *yhat*:

```
. replace yhat = exp(yhat)
(71 real changes made)
```

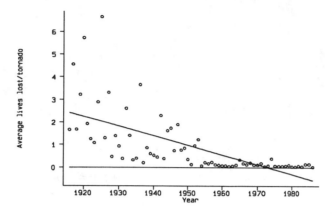

**Figure 9.4**

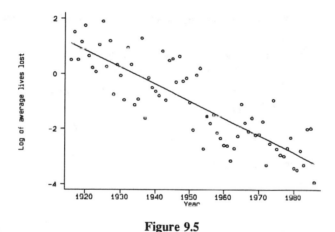

**Figure 9.5**

Graphing these inverse-transformed predicted values reveals the regression curve (Figure 9.6):

```
. graph avlost yhat year, connect(.s) symbol(Oi) ylabel(0,1,2,3,4,5,6)
     xlabel(1920,1930,1940,1950,1960,1970,1980) yline(0)
```

Contrast Figures 9.5–9.6 with Figure 9.4 to see how logarithms made the analysis both simpler and more realistic.

For a multiple-regression example, we return to the *nations.dta* analysis of Chapter 8. Figure 8.1 showed the inappropriateness of straight-line models with these data. Furthermore, the skew of *gnpcap* and *chldmort* may cause leverage and influence problems. Experimenting with power transformations reveals that the log of *gnpcap* and square root of mortality rate are more symmetrical than the original variables:

```
. use c:\stustata\nations, clear
. generate lognp = ln(gnpcap)/ln(10)
. label variable lognp "log-10 of per cap GNP"
. generate srmort = sqrt(chldmort)
. label variable srmort "square root child mortality"
```

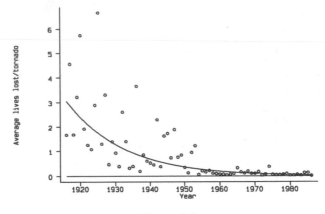

**Figure 9.6**

Univariate effects of these transformations show in graphs of transformed versus raw variables:

```
. graph lognp gnpcap, ylabel xlabel oneway twoway box
. graph srmort chldmort, ylabel xlabel oneway twoway box
```

Figure 9.7 combines both of these graphs, employing Professional (but not Student) Stata's ability to assemble new graphs by recycling other previously saved graphs:

```
. graph using figure7a figure7b, margin(10) t1(Effects of power
        transformations) saving(figure7)
```

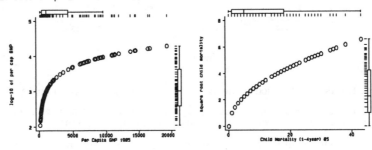

**Figure 9.7**

Logarithms are a stronger transformation than square roots, and useful here because *gnpcap* is more skewed than *chldmort*.

A scatterplot matrix (available under both Professional and Student Stata) shows that the transformations improved linearity (Figure 9.8; compare with Figure 8.1):

```
. graph lognp srmort birth, matrix half label
```

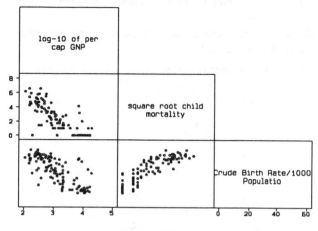

**Figure 9.8**

Since relations are now approximately linear, linear regression is appropriate:

```
. regress birth lognp srmort
(obs=109)
```

| Source | SS | df | MS |
|---|---|---|---|
| Model | 15837.9603 | 2 | 7918.98014 |
| Residual | 4238.1865 | 106 | 39.9828915 |
| Total | 20076.1468 | 108 | 185.890248 |

| | |
|---|---|
| Number of obs = | 109 |
| F( 2, 106) = | 198.06 |
| Prob > F = | 0.0000 |
| R-square = | 0.7889 |
| Adj R-square = | 0.7849 |
| Root MSE = | 6.3232 |

| Variable | Coefficient | Std. Error | t | Prob > |t| | Mean |
|---|---|---|---|---|---|
| birth | | | | | 32.78899 |
| lognp | -2.353738 | 1.686255 | -1.396 | 0.166 | 3.089199 |
| srmort | 5.577359 | .533567 | 10.453 | 0.000 | 2.485993 |
| _cons | 26.19488 | 6.362687 | 4.117 | 0.000 | 1 |

Unlike the raw-data regression, this transformed-variables version finds that per capita gross national product does not significantly affect birth rate, once we control for child mortality. The transformed-variables regression fits slightly better: $R_a^2$ = .7849 instead of .6715 (we can compare $R_a^2$ here because the $Y$ variable was not transformed). Leverage plots would show that the curvilinearity evident in the raw-data regression of Chapter 8 is much reduced in this transformed-variables regression.

## Conditional Effect Plots

Conditional effect plots trace the predicted values of $Y$ as a function of one $X$ variable, with the other $X$ variables held constant at arbitrary values such as means, medians, or quartiles. These plots help visualize the implications of transformed-variables regressions.

An equation for predicted birth rates as a function of *lognp*, with *srmort* held at its mean:

```
. generate yhat1 = _b[_cons]+_b[lognp]*lognp+_b[srmort]*2.485993
. label variable yhat1 "cond. effect lognp, srmort=mean"
```

The  `_b[varname]`  terms refer to the regression coefficient on *varname* from the Stata session's most recent regression.

For a conditional effect plot, graph *yhat1* (after inverse transformation if needed) against the untransformed *X* variable (Figure 9.9):

```
. graph yhat1 gnpcap, connect(s) symbol(i) ylabel xlabel l1(Birth Rate)
```

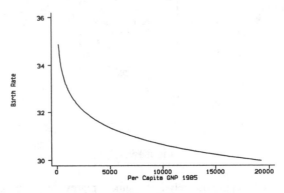

**Figure 9.9**

Similarly, to get predicted birth rates as a function of *srmort*, with *lognp* held at its mean:

```
. generate yhat2 = _b[_cons]+_b[lognp]*3.089199+_b[srmort]*srmort
. label variable yhat2 "cond. effect srmort, lognp=mean"
```

To graph this relation (Figure 9.10):

```
. graph yhat2 chldmort, connect(s) symbol(i) ylabel xlabel l1(Birth Rate)
```

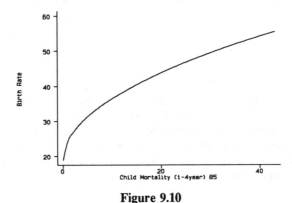

**Figure 9.10**

Figure 9.10 seems more plausible than the straight-line fit we imposed in Chapter 8.

How can we compare the strength of different *X* variables' effects?  One possibility is to draw conditional effect plots with identical *Y* scales, and vertical lines marking the 10th and 90th percentiles of the *X*-variable distribution (found by  **summarize, detail**).  The vertical distance traveled by the regression curve over the middle 80% of *X* values provides a visual comparison of effect magnitude.  For example, we might create these two separate plots using **yscale()** to control vertical scales and **xline()** to draw 10th and 80th-percentile lines:

```
. graph yhat1 gnpcap, connect(s) symbol(i) ylabel xlabel ll(Birth Rate)
      yscale(20,60) xline(230,10890) saving(fig11a)
. graph yhat2 chldmort, connect(s) symbol(i) ylabel xlabel ll(Birth Rate)
      yscale(20,60) xline(0,27) saving(fig11b)
```

Figure 9.11 combines both plots into a single image (using Stage), dramatizing the much stronger effect of mortality rate—as separate graphs (Figures 9.9 and 9.10) did not.

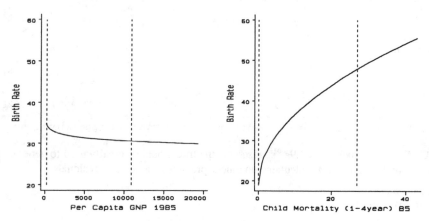

**Figure 9.11**

## Nonlinear Regression Example 1

Variable transformations allow fitting some curvilinear relations using the familiar techniques of intrinsically linear models. Intrinsically nonlinear models, in contrast, require a different class of fitting techniques. The **nl** command performs nonlinear regression by iterative least squares (Royston 1992; Danuso 1991). This section introduces **nl** using a dataset of simple examples.

```
. use c:\stustata\nonlin, clear
(artificial data--nonlinear)
. describe
```

```
Contains data from c:\stustata\nonlin.dta
  Obs:    100 (max=  2620)              Artificial data--nonlinear
  Vars:     6 (max=    99)
Width:     24 (max=   200)
  1. x               float   %9.0g     X variable
  2. y1              float   %9.0g     exponential growth model
  3. y2              float   %9.0g     negative exponential model
  4. y3              float   %9.0g     two-term exponential model
  5. y4              float   %9.0g     logistic growth model
  6. y5              float   %9.0g     Gompertz growth model
Sorted by:
```

The *nonlin.dta* data are manufactured; I defined variables *y1–y5* from various nonlinear functions of *x*, plus random error. *y1*, for example, approximately follows an exponential growth model. To estimate the parameters of this growth model, type:

```
. nl expo y1 x
(obs = 100)
Iteration 0:   residual SS =   17655.86
Iteration 1:   residual SS =   2093.249
```

```
Iteration 2:    residual SS =     1642.24
Iteration 3:    residual SS =    538.3342
Iteration 4:    residual SS =    439.5422
Iteration 5:    residual SS =    439.3995
Iteration 6:    residual SS =    439.3995
```

```
    Source |        SS        df        MS              Number of obs =        100
---------+------------------------------             F(  2,    98) =    2461.82
    Model |   22075.9416      2   11037.9708            Prob > F      =     0.0000
 Residual |   439.399477     98   4.48366813            R-square      =     0.9805
---------+------------------------------             Adj R-square  =     0.9801
    Total |   22515.3411    100   225.153411            Root MSE      =    2.117467
                                                      Res. dev.     =   431.8116
```

Exponential Curve
y1=B1*exp(B2*x)

```
------------------------------------------------------------------------
     y1 |     Coef.    Std. Err.       t      P>|t|    [95% Conf. Interval]
--------+---------------------------------------------------------------
     B1 |   3.904002    .216591     18.025    0.000     3.474184     4.33382
     B2 |   .0203735   .0006901     29.521    0.000      .019004    .0217431
------------------------------------------------------------------------
```

(SE's, P values, CI's, and correlations are asymptotic approximations)

**nl**'s final estimates, $y1 = 3.9e^{.02x}$, resemble the true model originally used to generate these data: $y1 = 4e^{.02x}$. **nlpred** calculates nonlinear predicted values and residuals:

```
. nlpred yhat1
. nlpred e1, resid
```

Figure 9.12 shows the good fit ($R^2 = .98$) between model and data:

```
. graph y1 yhat1 x, symbol(Oi) connect(.s) ylabel xlabel
```

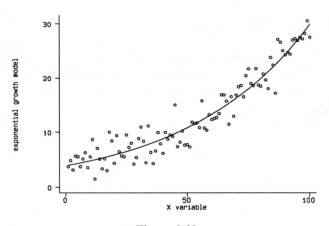

**Figure 9.12**

The **expo** part of our **nl expo y1 x** command specified an exponential growth model by calling a brief program named nlexpo.ado. Student Stata includes several such programs:

**nl expo**    exponential growth model (*nonlin.dta* example: **nl expo y1 x**)
**nl negexp**  negative exponential growth model (*nonlin.dta* example: **nl negexp y2 x**)
**nl 2exp**    2-term exponential growth model (*nonlin.dta* example: **nl 2exp y3 x**)
**nl logis**   logistic growth model (*nonlin.dta* example: **nl logis y4 x**)
**nl gomp**    Gompertz growth model (*nonlin.dta* example: **nl gomp y5 x**)

Users can easily write additional nl*function* programs of their own. Here, in its entirety, is the nlexpo.ado program for an exponential growth model:

```
program define nlexpo            /*   Exponential Curve    */
    if "'1'"=="?" {
            mac def S_1 "B1 B2"
            mac def B1=1
            mac def B2=.01
            mac def S_2 "Exponential Curve"
            mac def S_3 "$S_E_depv=B1*exp(B2*'2')"
            exit
    }
    replace '1'=$B1*exp($B2*'2')
end
```

Examine `nlnegexp.ado`, `nl2exp.ado`, etc. for other examples, and type **help nl** for a full explanation. If you do not have Student Stata or its ado-files, you will need to type in `nlexpo.ado` in order for commands like **nl expo y1 x** to work.

## Nonlinear Regression Example 2

Our second example involves real data, and illustrates some tricks that may help in research. Dataset *lichen.dta* contains measurements of lichen growth observed on the Arctic island of Spitsbergen (from Werner 1990). These slow-growing symbionts are often used to date rock monuments and other deposits, so their growth rates interest scientists in several fields.

```
. use c:\stustata\lichen, clear
(Lichen Growth--Werner (1990))
. describe

Contains data from c:\stustata\lichen.dta
  Obs:     11 (max=  2620)              Lichen Growth--Werner (1990)
  Vars:     8 (max=    99)
Width:     88 (max=   200)
    1. locale      str31    %31s        Locality and feature
    2. point       str1     %9s         Control point
    3. date        int      %8.0g       Date
    4. age         int      %8.0g       Age in years
    5. rshort      float    %9.0g       Rhizocarpon short axis mm
    6. rlong       float    %9.0g       Rhizocarpon long axis mm
    7. pshort      int      %8.0g       P.minuscula short axis mm
    8. plong       int      %8.0g       P.minuscula long axis mm
Sorted by:
```

Lichens characteristically exhibit a period of relatively fast early growth, gradually slowing, as suggested by the lowess-smoothed curve in Figure 9.13:

```
. ksm rlong age, lowess ylabel xlabel
```

Lichenometricians seek to summarize and compare such patterns by drawing growth curves. Their growth curves typically do not employ an explicit mathematical model, but we can fit one here to illustrate the process of nonlinear regression.

Gompertz curves are asymmetrical S-curves, widely used to model biological growth:

$$Y_i = B_1 \times \exp(-B_2 \times \exp(-B_3 X_i)) + \varepsilon_i \qquad [9.1]$$

Equation [9.1] might be a reasonable model for lichen growth too.

Especially when working with sparse data and/or a relatively complex model, nonlinear regression programs often prove to be very sensitive to their initial parameter estimates. Given unreasonable initial values, the iterations may never converge on a reasonable solution. But what values for $B_1$, $B_2$, and $B_3$ in [9.1] are reasonable, to fit the data of Figure 9.13?

Previous experience with similar data, or publications by other researchers, could supply suitable initial values. Alternatively, we can estimate parameters by trial and error using **graph**.

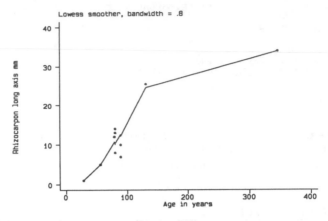

**Figure 9.13**

The Gompertz $B_1$ parameter determines the curve's asymptotic upper limit. $B_2$ determines the curve's takeoff point (higher values move it rightwards), and $B_3$ determines how steeply it climbs. As a wild guess, we might try generating predicted $Y$ values based on a Gompertz curve with parameters $B_1 = 34$, $B_2 = 50$, and $B_3 = .1$:

```
. generate yhat = 34*exp(-50*exp(-.1*age))
```

To see the results (top left in Figure 9.14) type:

```
. graph rlong yhat age, connect(.1) sort symbol(Si) saving(fig9_14a)
        title("yhat = 34*exp(-50*exp(-.1*age))")
```

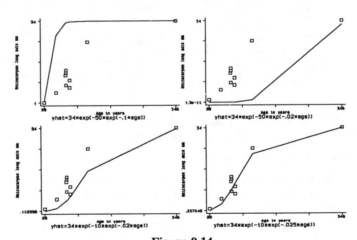

**Figure 9.14**

This first attempt climbs too steeply. Using <PgUp> and **replace**, next try:

```
. replace yhat = 34*exp(-50*exp(-.02*age))
. graph rlong yhat age, connect(.1) sort symbol(Si) saving(fig9_14b)
        title("yhat=34*exp(-50*exp(-.02*age))")
```

The results appear at top right in Figure 9.14. Now the curve is not steep enough. After several more attempts, we might settle for the version at lower right in Figure 9.14:

```
. replace yhat = 34*exp(-10*exp(-.025*age))
```

Thus we acquire some "ballpark" initial values:  $B_1 = 34$ ,  $B_2 = 10$ , and  $B_3 = .025$ .

If we intend eventually to graph our nonlinear model, we should make sure the data contain a good range of closely spaced $X$ values. Actual ages of the 11 lichen samples in the data range from 28 to 346 years. We can create 89 additional artificial cases, with "ages" from 0 to 352 in 4-year increments, by the following commands:

```
. range newage 0 396 100
. replace age = newage[_n-11] if age==.
```

The first command created a new variable, *newage*, with 100 values ranging from 0 to 396 in 4-year increments. In so doing, we also created 89 new artificial cases, with missing values on all variables but *newage*. The **replace** command replaces the missing artificial-case $x$ values with *newage*, starting at 0. The first 15 cases in our data now look like this:

```
. list rlong age newage in 1/15

             rlong         age      newage
  1.             1          28           0
  2.             5          56           4
  3.            12          79           8
  4.            14          80          12
  5.            13          80          16
  6.             8          80          20
  7.             7          89          24
  8.            10          89          28
  9.            34         346          32
 10.            34         346          36
 11.          25.5         131          40
 12.             .           0          44
 13.             .           4          48
 14.             .           8          52
 15.             .          12          56
```

```
. summ rlong age newage

Variable |     Obs        Mean    Std. Dev.       Min         Max
---------+-----------------------------------------------------------
   rlong |      11    14.86364    11.31391         1          34
     age |     100      170.68    104.7042         0         352
  newage |     100         198    116.046          0         396
```

We could **drop newage**. Only the original 11 cases have nonmissing *rlong* values, so only they will be used in estimation. Stata can calculate predicted values for all 100 "cases" with nonmissing *age* values, however—thereby allowing us to accurately graph the smooth curve, instead of line-segment approximations as seen in Figure 9.14.

We can now use **nl** to fit a Gompertz curve to the lichen measurement *rlong* by issuing the command **nl gomp**, followed by an **init()** option specifying the initial values guessed earlier from Figure 9.14.

```
. nl gomp rlong age, init(B1=34,B2=10,B3=.025)
(obs = 11)

Iteration 0:    residual SS =    90.79622
Iteration 1:    residual SS =    88.17865
Iteration 2:    residual SS =    77.15453
Iteration 3:    residual SS =    77.09037
Iteration 4:    residual SS =    77.08889
Iteration 5:    residual SS =    77.08888
Iteration 6:    residual SS =    77.08888
```

```
    Source |       SS        df       MS              Number of obs =         11
-----------+------------------------------           F(  3,    8) =      125.68
     Model |  3633.16112      3  1211.05371           Prob > F     =      0.0000
  Residual |  77.0888815      8  9.63611018           R-square     =      0.9792
-----------+------------------------------           Adj R-square =      0.9714
     Total |    3710.25      11  337.295455           Root MSE     =    3.104208
                                                      Res. dev.    =    52.63435
Gompertz Curve
rlong = B1*exp(-B2*exp(-B3*age))
------------------------------------------------------------------------------
     rlong |    Coef.     Std. Err.       t     P>|t|      [95% Conf. Interval]
-----------+------------------------------------------------------------------
        B1 |   34.3664    2.267169     15.158    0.000      29.1383     39.5945
        B2 |  6.909933    3.456772      1.999    0.081     -1.061397    14.88126
        B3 |  .0217683    .0060822      3.579    0.007      .0077426    .0357939
------------------------------------------------------------------------------
```

(SE's, P values, CI's, and correlations are asymptotic approximations)

The 95% confidence intervals indicate that B1 differs significantly from 1, but B2 does not; perhaps a simpler model would fit equally well.

Obtain predicted values as before, with the **nlpred** command:

```
. drop yhat
. nlpred yhat
. nlpred resid, resid
```

Figure 9.15 graphs the fit between predicted values (*yhat*) and data:

```
. graph rlong yhat age, connect(.s) symbol(Oi) ylabel(0,10,20,30)
        xlabel(0,100,200,300)
```

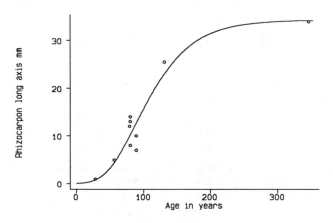

**Figure 9.15**

## Also Type help

| | |
|---|---|
| **generate** | compute new variable from algebraic expression |
| **graph** | graphs including median band regression and scatterplot matrices |
| **ksm** | lowess smoothing |
| **ladder** | search ladder of powers for normalizing transformation |
| **nl** | nonlinear regression |
| **range** | create variable with values over a specified range |
| **regress** | linear regression |

# 10
# Robust Regression

Stata's basic **regress**, **fit**, **oneway**, **anova**, and **ttest** commands, among others, perform ordinary least squares (OLS) regression. The popularity of OLS derives partly from its theoretical advantages given "ideal" data. If errors are normally, independently, and identically distributed (normal *i.i.d.*), then OLS is more efficient than any other unbiased estimator. The flip side of this statement often gets overlooked: if errors are not normal, or not *i.i.d.*, then other unbiased estimators may outperform OLS. In fact, the efficiency of OLS degrades quickly in the face of heavy-tailed (outlier-prone) error distributions. Yet such distributions are common in many fields.

OLS tends to track outliers, fitting them at the expense of the rest of the sample. Over the long run, this leads to greater sample-to-sample variation or inefficiency when samples often contain outliers. Robust regression methods aim to achieve almost the efficiency of OLS with ideal data, and substantially better-than-OLS efficiency with non-ideal (for example, not normal *i.i.d.*) situations. "Robust regression" encompasses a variety of different techniques, each with advantages and drawbacks for dealing with problematic data.

This chapter introduces three robust techniques, briefly comparing them to each other and OLS. Chapter 14 tells more about Monte Carlo research, which plays a central role in evaluating robustness.

## Regression with Ideal Data

To clarify the issue of robustness, we will explore the contrived dataset *robust.dta*:

```
Contains data from c:\stustata\robust.dta
  Obs:    20 (max=  2620)              Artificial data--robustness
  Vars:   10 (max=    99)
 Width:   40 (max=   200)
   1. x              float   %4.2f      Normal X
   2. e1             float   %4.2f      Normal errors
   3. y1             float   %4.2f      y1 = 10 + 2*x + e1
   4. e2             float   %4.2f      Normal errors with 1 outlier
   5. y2             float   %4.2f      y2 = 10 + 2*x + e2
   6. x3             float   %4.2f      Normal X with 1 leverage case
   7. e3             float   %4.2f      Normal errors with 1 extreme
   8. y3             float   %4.2f      y3 = 10 + 2*x3 + e3
   9. e4             float   %4.2f      Skewed errors
  10. y4             float   %4.2f      y4 = 10 + 2*x + e4
Sorted by:
```

The variables *x* and *e1* each contain 20 random values from a standard normal distribution. *y1* contains 20 values produced by the regression model $y1 = 10 + 2x + e1$. The commands that manufactured these first three variables are:

```
. clear
. set obs 20
. generate x = invnorm(uniform())
. generate e1 = invnorm(uniform())
. generate y1 = 10 + 2*x + e1
```

With real data, coding mistakes and other aberrations sometimes create wild errors. To simulate this, we might shift the second case's error from −0.89 to 19.89:

```
. generate e2 = e1
. replace e2 = 19.89 in 2
. generate y2 = 10 + 2*x + e2
```

Similar manipulations produce the other variables in *robust.dta*.

   *y1* and *x* present an ideal regression problem: the expected value of *y1* really is a linear function of *x* (E[*y1*] = 10 + 2*x*), and errors come from normal, independent, and identical distributions—because we defined them that way. OLS does a good job of estimating the true intercept (10) and slope (2), obtaining the line shown in Figure 10.1:

```
. regress y1 x
```

| Source | SS | df | MS |       |        | Number of obs | = | 20 |
|--------|----|----|----|-------|--------|---------------|---|----|
|        |    |    |    |       |        | F( 1, 18) | = | 108.25 |
| Model | 134.059351 | 1 | 134.059351 |  |  | Prob > F | = | 0.0000 |
| Residual | 22.29157 | 18 | 1.23842055 |  |  | R-square | = | 0.8574 |
|        |    |    |    |       |        | Adj R-square | = | 0.8495 |
| Total | 156.350921 | 19 | 8.22899586 |  |  | Root MSE | = | 1.1128 |

| y1 | Coef. | Std. Err. | t | P>|t| | [95% Conf. Interval] |
|----|-------|-----------|---|-------|----------------------|
| x | 2.048057 | .1968465 | 10.404 | 0.000 | 1.634498    2.461616 |
| _cons | 9.963161 | .2499861 | 39.855 | 0.000 | 9.43796    10.48836 |

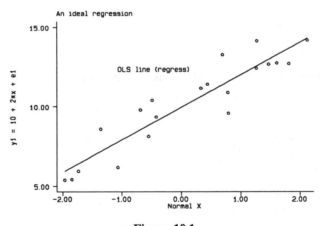

**Figure 10.1**

   An iteratively reweighted least squares (IRLS) procedure, **rreg**, obtains robust regression estimates. On each iteration, **rreg** estimates regression parameters, calculates the residuals, and downweights cases with large residuals. The process repeats until weights no longer show much change:

```
. rreg y1 x

    Huber iteration 1:  maximum difference in weights = .3577441
    Huber iteration 2:  maximum difference in weights = .02181578
 Biweight iteration 3:  maximum difference in weights = .14421374
 Biweight iteration 4:  maximum difference in weights = .01320279
 Biweight iteration 5:  maximum difference in weights = .00265408

Robust regression estimates                     Number of obs =       20
                                                  F( 1,    18) =    79.96
                                                  Prob > F      =   0.0000

------------------------------------------------------------------------
     y1 |      Coef.   Std. Err.        t     P>|t|    [95% Conf. Interval]
--------+---------------------------------------------------------------
      x |   2.047813   .2290049     8.942    0.000     1.566692    2.528935
  _cons |   9.936164   .2908259    34.165    0.000     9.325161    10.54717
------------------------------------------------------------------------
```

**rreg** employs a combination of Huber and biweight functions (described in *Regression with Graphics*). This sample's *y1* distribution includes no serious outliers, however, so here **rreg** is unneeded. **rreg**'s intercept and slope estimates resemble those obtained by **regress** (and aren't far from the true values 10 and 2), but have slightly larger standard errors. Given normal *i.i.d.* errors, as in this example, **rreg** generally possesses about 95% of the efficiency of OLS.

   **rreg** and **regress** both belong to the family of *M*-estimators (for maximum-likelihood). An alternative estimation strategy called *L*-estimation fits quantiles of *y*, rather than its expectation or mean. For example, we could model how the median (.5 quantile) of *y* changes with *x*. **qreg**, an *L1*-type estimator, accomplishes such quantile regression and provides another robust alternative to **regress**:

```
. qreg y1 x
Iteration  1:  WLS sum of weighted deviations =  17.711531

Iteration  1: sum of abs. weighted deviations =  17.130001
Iteration  2: sum of abs. weighted deviations =  16.858602

Median Regression                               Number of obs -       20
   Raw sum of deviations    46.84 (about 10.4)
   Min sum of deviations  16.8586               Pseudo R2     =   0.6401

------------------------------------------------------------------------
     y1 |      Coef.   Std. Err.        t     P>|t|    [95% Conf. Interval]
--------+---------------------------------------------------------------
      x |   2.139896   .2590447     8.261    0.000     1.595664    2.684129
  _cons |   9.65342    .3564108    27.085    0.000     8.904628    10.40221
------------------------------------------------------------------------
```

Although **qreg** obtains reasonable parameter estimates, its standard errors here exceed those of **regress** (OLS) and **rreg**. Given ideal data, **qreg** is the least efficient of these three estimators. The following sections view their performance with less ideal data.

## *Y*-Outliers

The variable *y2* is identical to *y1*, but with one outlier due to the "wild" error of case #2. OLS has little resistance to outliers, so this shift in case #2 (at upper left in Figure 10.2) substantially changes the **regress** results:

```
. regress y2 x

  Source |      SS        df       MS                Number of obs =      20
---------+---------------------------------          F(  1,    18) =    0.97
   Model |   18.764271     1    18.764271            Prob > F      =  0.3378
Residual |  348.233471    18    19.3463039           R-square      =  0.0511
---------+---------------------------------          Adj R-square  = -0.0016
   Total |  366.997742    19    19.3156706           Root MSE      =  4.3984


      y2 |     Coef.   Std. Err.        t     P>|t|      [95% Conf. Interval]
---------+----------------------------------------------------------------------
       x |   .7662304   .7780232     0.985    0.338      -.8683356    2.400796
   _cons |   11.1579    .9880542    11.293    0.000       9.082078   13.23373
```

The outlier raises the OLS intercept (from 9.96 to 11.16) and lessens the slope (from 2.05 to 0.77). Standard errors quadruple, and the OLS slope (solid line in Figure 10.2) no longer significantly differs from zero.

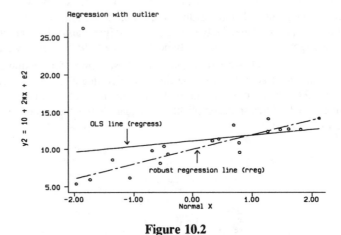

**Figure 10.2**

The outlier has little impact on **rreg**, however. Its coefficients barely change, and remain close to the true parameters 10 and 2. Nor do the **rreg** standard errors increase much.

```
. rreg y2 x, genwt(weight)

   Huber iteration 1:   maximum difference in weights = .32200664
   Huber iteration 2:   maximum difference in weights = .27149647
   Huber iteration 3:   maximum difference in weights = .01953268
Biweight iteration 4:   maximum difference in weights = .14250112
Biweight iteration 5:   maximum difference in weights = .11161703
Biweight iteration 6:   maximum difference in weights = .01596266
Biweight iteration 7:   maximum difference in weights = .00201833

Robust regression estimates                         Number of obs =      20
                                                    F(  1,    18) =   72.82
                                                    Prob > F      =  0.0000


      y2 |     Coef.   Std. Err.        t     P>|t|      [95% Conf. Interval]
---------+----------------------------------------------------------------------
       x |   1.981267   .2321768     8.533    0.000       1.493482    2.469052
   _cons |   10.01053   .2948541    33.951    0.000       9.391065    10.63
```

The **genwt(***weight***)** option told **rreg** to save robust weights as a variable named *weight*:

```
. predict resid, resid
. list y2 x resid weight

        y2      x       resid       weight
 1.   5.37   -1.97   -.7374348     .9557462
 2.  26.19   -1.85    19.84481            0
 3.   5.93   -1.74   -.6331263     .9672729
 4.   8.58   -1.36    1.263992     .8732737
 5.   6.16   -1.07   -1.730575     .7690216
 6.   9.80   -0.69    1.156544     .8933188
 7.   8.12   -0.55   -.8008341     .9479429
 8.  10.40   -0.49     1.36029      .853985
 9.   9.35   -0.42    .1716017     .9976033
10.  11.16    0.33    .4956509     .9799808
11.  11.40    0.44    .5177113      .978166
12.  13.26    0.69    1.882395     .7303757
13.  10.88    0.78   -.6759189     .9627756
14.   9.58    0.79   -1.995732      .699545
15.  12.41    1.26   -.0969274     .9992201
16.  14.14    1.27     1.61326      .797848
17.  12.66    1.47   -.2629935     .9943067
18.  12.74    1.61    -.460371     .9826358
19.  12.70    1.81   -.8966242     .9350107
20.  14.19    2.12   -.0208171     .9999629
```

Figure 10.3 graphs robust weights against residuals, outlining part of the final biweight function.

```
. graph weight resid, yline(0,1) xline(0) b2(Residual) xlabel ylabel noaxis
```

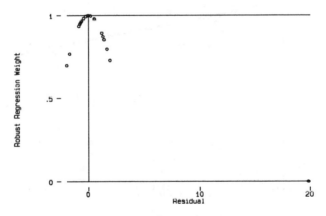

**Figure 10.3**

Residuals near zero produce weights near one; farther-out residuals get progressively lower weights. Case #2 receives zero weight, and so has no effect on the final **rreg** estimates. Those final estimates (but <u>not</u> the correct standard errors or tests) could be duplicated by this weighted regression (results not shown):

```
. regress y2 x [aweight=weight]
```

Applied to the regression of *y2* on *x*, **qreg** also resists the outlier's influence and performs much better than **regress**—but not as well as **rreg**. **qreg** remains less efficient than **rreg** (larger standard errors), and its estimates are slightly farther from 10 and 2.

## X-Outliers (Leverage)

**rreg** and **qreg** deal comfortably with *Y*-outliers, unless the cases with unusual *Y* values have unusual *X* values (leverage) too. The *y3* and *x3* variables in *robust.dta* present an extreme example of leverage. Apart from the leverage case (#2), these variables equal *y1* and *x*.

Case #2's high leverage, combined with its exceptional *y3* value, make it influential: **regress**, **rreg**, and **qreg** all track this outlier, reporting that the "best-fitting" line has a negative slope (Figure 10.4). For example:

```
. regress y3 x3

  Source |       SS       df       MS                  Number of obs =      20
---------+------------------------------              F(  1,    18) =   11.01
   Model | 139.306724     1   139.306724              Prob > F      =  0.0038
Residual | 227.691018    18   12.649501              R-square      =  0.3796
---------+------------------------------              Adj R-square  =  0.3451
   Total | 366.997742    19   19.3156706              Root MSE      =  3.5566


      y3 |    Coef.   Std. Err.      t     P>|t|      [95% Conf. Interval]
---------+--------------------------------------------------------------------
      x3 | -.6212248   .1871973    -3.319   0.004    -1.014512   -.227938
   _cons | 10.80931    .8063436    13.405   0.000     9.115244   12.50337
```

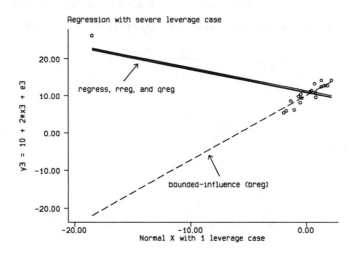

**Figure 10.4**

Figure 10.4 illustrates the fact that **rreg** and **qreg** are not robust against leverage (*X*-outliers). For robustness against leverage, we turn to yet another estimator, **breg**, that performs a rough form of bounded-influence regression (described in Hamilton 1991a, 1992a). **breg** downweights *Y*-outlier cases in the same manner as **rreg**, but unlike **rreg**, **breg** downweights *X*-outliers (relatively high-leverage cases) as well. The standard errors and tests **breg** prints are inaccurate, but it otherwise succeeds where **regress**, **rreg**, and **qreg** fail: at discerning the up-to-right (*y3* = 10 + 2*x3* + *e3*) pattern followed by 95% of the data.

```
. breg y3 x3
Huber iteration 1:   maximum difference in weights = .29473633
Huber iteration 2:   maximum difference in weights = .03001487
Huber iteration 3:   maximum difference in weights = .00345618
Biweight iteration 4:  maximum difference in weights = .99479991
```

|              |          |           |        | Number of obs = |      20 |
|---|---|---|---|---|---|
| y3 \| | Coef. | Std. Err. | t | P>\|t\| | [95% Conf. Interval] |
| x3 \| | 1.735501 | .2689594 | 6.453 | 0.000 | 1.170438 | 2.300563 |
| _cons \| | 10.13316 | .3318678 | 30.534 | 0.000 | 9.435931 | 10.83039 |

NOTE:  Standard errors and tests are not correct

Diagnostic statistics like hat diagonals, Cook's *D*, or *DFBETAS* (Chapter 8) can help analysts detect, and possibly exclude, excessively influential cases like those in Figures 10.2–10.4. Outlier deletion is an all-or-nothing decision, however. Smooth downweighting, as in **rreg** or **breg**, provides a less arbitrary and generally more efficient way to cope with outliers. "Outlierhood" is a matter of degree, and in practice it is often less clear-cut than in Figures 10.2 or 10.4.

The examples in Figures 10.2–10.4 involve single outliers, but robust procedures should be able to handle more. Monte Carlo research shows that estimators like **rreg**, **qreg**, and **breg** generally remain unbiased, with better-than-OLS efficiency, when applied to heavy-tailed (outlier-prone) but symmetrical error distributions. The next section illustrates what can happen when the error distribution is not symmetrical.

## Asymmetrical Error Distributions

The variable *e4* in *robust.dta* has a skewed and outlier-filled distribution: *e4* equals *e1* (a standard normal variable) raised to the fourth power, then adjusted to have 0 mean. These skewed errors, plus the linear relation with *x*, define the last *Y* variable: $y4 = 10 + 2x + e4$. Regardless of an error distribution's shape, OLS remains an unbiased estimator—over the long run, its estimates should center on the true parameter values:

```
. regress y4 x
```

| Source \| | SS | df | MS | | Number of obs = | 20 |
|---|---|---|---|---|---|---|
| Model \| | 155.870383 | 1 | 155.870383 | | F( 1, 18) = | 6.97 |
| Residual \| | 402.341909 | 18 | 22.3523283 | | Prob > F = | 0.0166 |
| | | | | | R-square = | 0.2792 |
| | | | | | Adj R-square = | 0.2392 |
| Total \| | 558.212291 | 19 | 29.3795943 | | Root MSE = | 4.7278 |

| y4 \| | Coef. | Std. Err. | t | P>\|t\| | [95% Conf. Interval] |
|---|---|---|---|---|---|
| x \| | 2.208388 | .8362862 | 2.641 | 0.017 | .4514157 | 3.96536 |
| _cons \| | 9.975681 | 1.062046 | 9.393 | 0.000 | 7.744406 | 12.20696 |

The same is not true for most robust estimators. Unless errors are symmetrical, the median line sought by **qreg** does not theoretically coincide with the expected-*Y* line sought by **regress**. As long as the errors' skew reflects only a small fraction of their distribution, **rreg** or **breg** may exhibit little bias. But when the entire distribution is skewed, as with *e4*, **rreg** or **breg** will downweight mostly one side, resulting in noticeably biased *Y*-intercept estimates:

```
. rreg y4 x
     Huber iteration 1:    maximum difference in weights = .88476676
     Huber iteration 2:    maximum difference in weights = .75509983
     Huber iteration 3:    maximum difference in weights = .67451477
     Huber iteration 4:    maximum difference in weights = .18492019
     Huber iteration 5:    maximum difference in weights = .19552952
     Huber iteration 6:    maximum difference in weights = .10699838
     Huber iteration 7:    maximum difference in weights = .10186553
     Huber iteration 8:    maximum difference in weights = .04786515
  Biweight iteration 9:    maximum difference in weights = .30123091
  Biweight iteration 10:   maximum difference in weights = .37368807
  Biweight iteration 11:   maximum difference in weights = .1121695
  Biweight iteration 12:   maximum difference in weights = .01568955
  Biweight iteration 13:   maximum difference in weights = .00116086
```

```
Robust regression estimates                      Number of obs =        20
                                                  F( 1,   18) = 1394.78
                                                  Prob > F     =  0.0000

---------------------------------------------------------------------------
     y4 |     Coef.   Std. Err.       t    P>|t|     [95% Conf. Interval]
---------+-----------------------------------------------------------------
      x |   1.952073   .0522688    37.347   0.000     1.842261    2.061886
   _cons |   7.476669    .066379   112.636   0.000     7.337212    7.616127
---------------------------------------------------------------------------
```

Although **rreg**'s estimated $Y$-intercept in Figure 10.5 is too low, the slope remains parallel to the OLS line and the true model. In fact, being less affected by outliers, the **rreg** slope is closer to the true slope and obtains a much smaller standard error than **regress**. This illustrates the tradeoff of using **rreg** or similar estimators with skewed errors: we risk getting biased estimates of the $Y$-intercept, but can still expect unbiased and relatively precise estimates of other regression coefficients. In many applications such coefficients are substantively more interesting than the $Y$-intercept, so this tradeoff is worthwhile.

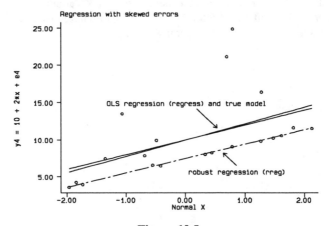

**Figure 10.5**

## Robust Analysis of Variance

**rreg** can also perform robust analysis of variance or analysis of covariance, once they are recast in regression form. For illustration, consider these data on college faculty salaries (complete $n$ = 226 data in *faculty1.dta*; with Student Stata use the $n$ = 150 random subsample in *faculty2.dta*):

```
. use c:\stustata\faculty1
. describe
```

```
Contains data from c:\stustata\faculty1.dta
  Obs:    226 (max=   2620)              College faculty salaries
  Vars:     6 (max=     99)
 Width:     9 (max=    200)
   1. rank          byte    %8.0g    rank   Academic rank
   2. sex           byte    %8.0g    sex    Gender (dummy variable)
   3. female        byte    %8.0g           Gender (effect coded)
   4. assoc         byte    %8.0g           Assoc Professor (effect coded)
   5. full          byte    %8.0g           Full Professor (effect coded)
   6. pay           float   %9.0g           1989-90 salary
Sorted by:
```

Faculty salaries increase with rank. Men have higher average salaries:

```
. tabulate sex rank, summ(pay) means
```

```
                        Means of 1989-90 salary

      Gender| Academic rank
      (dummy|
    variable)|    Assist       Assoc        Full        Total
  -----------+------------------------------------------+----------
        Male |     29280    38622.222   52084.902 | 41035.772
      Female | 28711.034    38019.048       47190 | 35228.052
  -----------+------------------------------------------+----------
       Total | 29022.188    38380.952   51569.649 | 39057.035
```

An ordinary (OLS) analysis of variance indicates that both *rank* and *sex* significantly affect salary. Their interaction is not significant:

```
. anova pay rank sex rank*sex
```

```
                         Number of obs =       226    R-square     =  0.7305
                         Root MSE      = 5108.21    Adj R square =  0.7244

         Source |  Partial SS    df       MS              F     Prob > F
    ------------+----------------------------------------------------------
          Model |  1.5560e+10     5   3.1120e+09         119.26    0.0000
                |
           rank |  7.6124e+09     2   3.8062e+09         145.87    0.0000
            sex |   127361829     1    127361829           4.88    0.0282
       rank*sex |  87997720.1     2   43998860.1           1.69    0.1876
                |
       Residual |  5.7406e+09   220   26093824.5
    ------------+----------------------------------------------------------
          Total |  2.1300e+10   225   94668810.3
```

But salary is not normally distributed, and the senior-rank averages reflect the influence of a few highly paid outliers. Suppose we want to check these results by performing a robust analysis of variance. We need effect-coded versions of the *rank* and *sex* variables:

```
. tab sex female
```

```
      Gender| Gender (effect coded)
      (dummy|
    variable)|        -1           1 |      Total
  -----------+----------------------+----------
        Male |       149           0 |        149
      Female |         0          77 |         77
  -----------+----------------------+----------
       Total|       149          77 |        226
```

```
. tab rank assoc

   Academic| Assoc Professor (effect coded)
      rank|        -1          0          1 |     Total
-----------+---------------------------------+----------
    Assist |        64          0          0 |        64
     Assoc |         0          0        105 |       105
      Full |         0         57          0 |        57
-----------+---------------------------------+----------
     Total|        64         57        105 |       226

. tab rank full

   Academic| Full Professor (effect coded)
      rank|        -1          0          1 |     Total
-----------+---------------------------------+----------
    Assist |        64          0          0 |        64
     Assoc |         0        105          0 |       105
      Full |         0          0         57 |        57
-----------+---------------------------------+----------
     Total|        64        105         57 |       226
```

*faculty1.dta* provides effect-coded variables (*female*, *assoc*, and *full*), but otherwise we could create them from *sex* and *rank* using a series of **generate** and **replace** statements. (See *Regression with Graphics* for more about effect coding.) We also need two interaction terms, representing female associate professors and female full professors:

```
. generate femassoc = female*assoc
. generate femfull = female*full
```

Males and assistant professors are "omitted categories" in this example.

Now we can duplicate the previous ANOVA using regression:

```
. regress pay assoc full female femassoc femfull

    Source |       SS       df       MS              Number of obs =     226
-----------+------------------------------           F(  5,   220) =  119.26
     Model | 1.5560e+10      5  3.1120e+09           Prob > F      =  0.0000
  Residual | 5.7406e+09    220  26093824.5           R-square      =  0.7305
-----------+------------------------------           Adj R-square  =  0.7244
     Total | 2.1300e+10    225  94668810.3           Root MSE      =  5108.2


       pay |      Coef.   Std. Err.       t     P>|t|    [95% Conf. Interval]
-----------+----------------------------------------------------------------
     assoc |  -663.8995   543.8499     -1.221   0.223   -1735.722    407.9229
      full |   10652.92   783.9227     13.589   0.000    9107.957    12197.88
    female |  -1011.174   457.6938     -2.209   0.028   -1913.199   -109.1483
  femassoc |   709.5864   543.8499      1.305   0.193    -362.236    1781.409
   femfull |  -1436.277   783.9227     -1.832   0.068   -2981.237    108.6819
     _cons |   38984.53   457.6938     85.176   0.000    38082.51    39886.56
```

```
. test assoc full

 ( 1)  assoc = 0.0
 ( 2)  full = 0.0

       F(  2,   220) =   145.87
            Prob > F =    0.0000
```

```
. test female

 ( 1)  female = 0.0

       F(  1,   220) =     4.88
            Prob > F =    0.0282
```

```
. test femassoc femfull

( 1)  femassoc = 0.0
( 2)  femfull = 0.0

       F(  2,   220) =     1.69
            Prob > F =     0.1876
```

**regress** followed by appropriate **test** commands obtains exactly the same $R^2$ and $F$ test results we found earlier using **anova**. Predicted values from this regression equal the mean salaries (compare with tabulation on page 133):

```
. predict predpay1
. label variable predpay1 "OLS predicted salary"
. tabulate sex rank, summ(predpay1) means
```

Means of OLS predicted salary

| Gender (dummy variable) | Academic rank | | | |
|---|---|---|---|---|
| | Assist | Assoc | Full | Total |
| Male | 29280 | 38622.223 | 52084.902 | 41035.772 |
| Female | 28711.035 | 38019.047 | 47190 | 35228.052 |
| Total | 29022.188 | 38380.952 | 51569.649 | 39057.036 |

Predicted values (means), $R^2$, and $F$ tests would also be the same regardless of which categories we chose to omit from the regression. Our "omitted categories," males and assistant professors, are not really absent. Their information is implied by the included categories: if a faculty member is not female, he must be male, and so forth.

To perform a robust analysis of variance, apply **rreg** to this model:

```
. rreg pay assoc full female femassoc femfull
    Huber iteration 1:  maximum difference in weights = .80164045
    Huber iteration 2:  maximum difference in weights = .17859399
    Huber iteration 3:  maximum difference in weights = .07641017
    Huber iteration 4:  maximum difference in weights = .017299
  Biweight iteration 5:  maximum difference in weights = .29500043
  Biweight iteration 6:  maximum difference in weights = .0541991
  Biweight iteration 7:  maximum difference in weights = .04360747
  Biweight iteration 8:  maximum difference in weights = .00833139
```

```
Robust regression estimates              Number of obs =     226
                                         F(  5,  220) = 138.29
                                         Prob > F     = 0.0000
```

| pay | Coef. | Std. Err. | t | P>\|t\| | [95% Conf. Interval] | |
|---|---|---|---|---|---|---|
| assoc | -315.6458 | 458.0815 | -0.689 | 0.492 | -1218.436 | 587.1439 |
| full | 9765.296 | 660.2935 | 14.789 | 0.000 | 8463.985 | 11066.61 |
| female | -749.4947 | 385.5128 | -1.944 | 0.053 | -1509.265 | 10.27604 |
| femassoc | 197.7841 | 458.0815 | 0.432 | 0.666 | -705.0056 | 1100.574 |
| femfull | -913.3491 | 660.2935 | -1.383 | 0.168 | -2214.659 | 387.961 |
| _cons | 38331.87 | 385.5128 | 99.431 | 0.000 | 37572.1 | 39091.64 |

```
. test assoc full

( 1)  assoc = 0.0
( 2)  full = 0.0

       F(  2,   220) =   182.73
            Prob > F =     0.0000
```

```
. test female

 ( 1)   female = 0.0

        F(  1,   220) =     3.78
              Prob > F =    0.0532

. test femassoc femfull

 ( 1)   femassoc = 0.0
 ( 2)   femfull = 0.0

        F(  2,   220) =     1.16
              Prob > F =    0.3143
```

**rreg** downweights several outliers, mainly high-paid male full professors. To see the robust means, again use predicted values:

```
. predict predpay2
. label variable predpay2 "Robust predicted salary"
. tabulate sex rank, summ(predpay2) means
```

                    Means of Robust predicted salary

| Gender (dummy variable) | Academic rank | | | |
|---|---|---|---|---|
| | Assist | Assoc | Full | Total |
| Male | 28916.148 | 38567.934 | 49760.008 | 40131.58 |
| Female | 28848.289 | 37464.512 | 46434.32 | 34918.387 |
| Total | 28885.4 | 38126.565 | 49409.935 | 38355.404 |

The male-female salary gap among assistant and full professors appears smaller if we use robust means. It does not entirely vanish, however, and the gender gap among associate professors slightly widens.


## Further Robust Applications

Diagnostic statistics and graphs (Chapter 8) and power transformations (Chapter 9) extend and safeguard the usefulness of **rreg**, as they do ordinary regression. **rreg** can also robustly perform simpler types of analysis. To obtain a 90% confidence interval for the mean of a single variable, $Y$, we could type either the usual confidence-interval command **ci**:

```
. ci Y, level(90)
```

or get the same interval by regression:

```
. regress Y, level(90)
```

For a robust 90% confidence interval, type:

```
. rreg Y, level(90)
```

In all three commands, the **level()** option specifies the desired degree of confidence. If we omit this option, Stata automatically displays a 95% confidence interval.

To compare two means, analysts typically employ a two-sample $t$ test (**ttest**) or oneway analysis of variance (**oneway** or **anova**). As seen earlier, we can perform equivalent tests (yielding identical $t$ and $F$ statistics) with regression, for example by regressing the measurement variable ($Y$) on a dummy variable (here called *group*) representing the two categories:

```
. regress Y group
```

A robust version of this test results from typing:

```
. rreg Y group
```

    **qreg** performs median regression by default, but it is actually a more general tool: it can fit a linear model for any quantile of $Y$, not just the median (.5 quantile). For example, commands like this can analyze how the first quartile (.25 quantile) of $Y$ changes with $X$:

```
. qreg Y X, quant(.25)
```

Assuming constant error variance, the slopes of the .25 and .75 quantile lines should be the same. **qreg** could thus perform a check for heteroscedasticity, or subtle kinds of nonlinearity.


## Also Type help

| | |
|---|---|
| **anova** | general analysis of variance/covariance |
| **breg** | bounded-influence robust regression (supplied with Student Stata) |
| **ci** | confidence intervals for means |
| **fit** | OLS with diagnostic statistics |
| **hreg** | OLS with jackknife (Huber) standard errors |
| **predict** | regression predicted values, residuals, diagnostics |
| **qreg** | quantile regression, including medians ($L1$-estimate) |
| **regress** | OLS regression; also WLS, 2SLS |
| **rreg** | robust regression ($M$-estimate) |
| **test** | test linear hypotheses after **regress**, **rreg**, etc. |
| **ttest** | one- and two-sample $t$ tests |
| **weight** | case weighting—analytical, frequency, sampling, importance |

# 11
# Logistic Regression

The regression and ANOVA methods described in Chapters 5–10 require measured $Y$ variables. Stata also offers a broad selection of techniques for other types of $Y$ variables, such as categorical, ordered-category, counted, and censored. Estimation commands include:

| | |
|---|---|
| **blogit** | logistic regression from blocked data (**help glogit**) |
| **bprobit** | probit regression from blocked data (**help glogit**) |
| **clogit** | conditional logistic regression |
| **cnreg** | censored-normal regression |
| **cox** | proportional hazards model |
| **glogit** | logistic regression from grouped data |
| **gprobit** | probit regression from grouped data (**help glogit**) |
| **logistic** | logistic regression, giving odds ratios and diagnostics |
| **logit** | logistic regression, giving estimated coefficients |
| **loglin** | loglinear modeling (supplied with Student Stata; see Judson 1992) |
| **mlogit** | multinomial logistic regression, with polytomous $Y$ variable |
| **ologit** | logistic regression with ordered-category $Y$ variable |
| **oprobit** | probit regression with ordered-category $Y$ variable (**help ologit**) |
| **poisson** | poisson regression, assuming Poisson-distributed $Y$ variable (counts) |
| **probit** | probit regression, with dichotomous $Y$ variable |
| **tobit** | tobit regression |

Various prediction, hypothesis-testing, graphing, and diagnostic commands, described in the appropriate **help** files, support these estimation commands. The estimation commands often have stepwise variants (page 169).

This chapter mainly focuses on the most popular categorical-$Y$ approach, called logistic or logit regression.[1] We begin with the simplest case, a dichotomous $Y$ variable.

## Space Shuttle Data

Our main example uses data from the first 25 flights of the U.S. space shuttle. These data contain evidence that, if properly analyzed, might have warned in advance that *Challenger* (the 25th flight, designated STS 51-L) should not be launched (*Report of the Presidential Commission on the Space Shuttle Challenger Accident* 1986).

---

[1]Some authors describe "logistic regression" and "logit regression" as different techniques. Others, including Stata and this book, follow the econometric tradition of treating "logit regression" and "logistic regression" as synonyms. Although Stata has separate **logit** and **logistic** commands, they differ only in output; both estimate the same models using maximum likelihood.

```
Contains data from c:\stustata\shuttle.dta
  Obs:     25 (max=  2620)                    Space Shuttle data
  Vars:     6 (max=    99)
 Width:     7 (max=   200)
    1. flight       byte    %8.0g    flbl     Flight
    2. month        byte    %8.0g             Month of Launch
    3. day          byte    %8.0g             Day of Launch
    4. year         int     %8.0g             Year of Launch
    5. distress     byte    %8.0g    dlbl     Thermal Distress Incidents
    6. temp         byte    %8.0g             Joint Temperature, degrees F
Sorted by:

. list
```

| | flight | month | day | year | distress | temp |
|---|---|---|---|---|---|---|
| 1. | STS-1 | 4 | 12 | 1981 | none | 66 |
| 2. | STS-2 | 11 | 12 | 1981 | 1 or 2 | 70 |
| 3. | STS-3 | 3 | 22 | 1982 | none | 69 |
| 4. | STS-4 | 6 | 27 | 1982 | . | 80 |
| 5. | STS-5 | 11 | 11 | 1982 | none | 68 |
| 6. | STS-6 | 4 | 4 | 1983 | 1 or 2 | 67 |
| 7. | STS-7 | 6 | 18 | 1983 | none | 72 |
| 8. | STS-8 | 8 | 30 | 1983 | none | 73 |
| 9. | STS-9 | 11 | 28 | 1983 | none | 70 |
| 10. | STS_41-B | 2 | 3 | 1984 | 1 or 2 | 57 |
| 11. | STS_41-C | 4 | 6 | 1984 | 3 plus | 63 |
| 12. | STS_41-D | 8 | 30 | 1984 | 3 plus | 70 |
| 13. | STS_41-G | 10 | 5 | 1984 | none | 78 |
| 14. | STS_51-A | 11 | 8 | 1984 | none | 67 |
| 15. | STS_51-C | 1 | 24 | 1985 | 3 plus | 53 |
| 16. | STS_51-D | 4 | 12 | 1985 | 3 plus | 67 |
| 17. | STS_51-B | 4 | 29 | 1985 | 3 plus | 75 |
| 18. | STS_51-G | 6 | 17 | 1985 | 3 plus | 70 |
| 19. | STS_51-F | 7 | 29 | 1985 | 1 or 2 | 81 |
| 20. | STS_51-I | 8 | 27 | 1985 | 1 or 2 | 76 |
| 21. | STS_51-J | 10 | 3 | 1985 | none | 79 |
| 22. | STS_61-A | 10 | 30 | 1985 | 3 plus | 75 |
| 23. | STS_61-B | 11 | 26 | 1985 | 1 or 2 | 76 |
| 24. | STS_61-C | 1 | 12 | 1986 | 3 plus | 58 |
| 25. | STS_51-L | 1 | 28 | 1986 | . | 31 |

This chapter studies three variables:

*distress*    the number of "thermal distress incidents," in which hot gas blow-through or charring damaged joint seals of a flight's booster rockets. Burn-through of a booster joint seal precipitated the *Challenger* disaster. Many previous flights had experienced similar but less severe damage, so the joint seals were known to be a source of possible danger.

*temp*    the calculated joint temperature at launch time, in degrees Fahrenheit. Temperature depends largely on weather. Rubber O-rings sealing the booster rocket joints become less flexible when cold.

*date*    date, measured in days elapsed since January 1, 1960 (an arbitrary starting point). *date* is generated from the month, day, and year of launch using the **mdytoe** (month-day-year to elapsed time; see **help date**) command:

```
. mdytoe month day year, generate(date)
. label variable date "Date (days since 1/1/60)"
```

    Launch date matters because several changes over the course of the shuttle program may have made it riskier. Booster rocket walls were thinned to save weight and increase payloads, and joint seals were subjected to higher-pressure testing. Furthermore, the reusable shuttle hardware was aging.  So we might ask, did the probability of booster joint damage (one or more distress incidents) increase with launch date?

*distress* is a labeled numeric variable:

```
. tabulate distress
```

| Thermal Distress Incidents | Freq. | Percent | Cum. |
|---|---|---|---|
| none | 9 | 39.13 | 39.13 |
| 1 or 2 | 6 | 26.09 | 65.22 |
| 3 plus | 8 | 34.78 | 100.00 |
| Total | 23 | 100.00 | |

Ordinarily, **tabulate** displays the labels, but the **nolabel** option reveals that the underlying numerical codes are 0="none", 1="1 or 2", and 2="3 plus":

```
. tabulate distress, nolabel
```

| Thermal Distress Incidents | Freq. | Percent | Cum. |
|---|---|---|---|
| 0 | 9 | 39.13 | 39.13 |
| 1 | 6 | 26.09 | 65.22 |
| 2 | 8 | 34.78 | 100.00 |
| Total | 23 | 100.00 | |

We can use these codes to create a new dummy variable, *any*, coded 0 for no distress and 1 for one or more distress incidents:

```
. generate any = distress
(2 missing values generated)
. replace any = 1 if distress==2
(8 real changes made)
. label variable any "Any Thermal Distress"
```

To see what this accomplished:

```
. tabulate distress any
```

| Thermal Distress Incidents | Any Thermal Distress | | Total |
|---|---|---|---|
| | 0 | 1 | |
| none | 9 | 0 | 9 |
| 1 or 2 | 0 | 6 | 6 |
| 3 plus | 0 | 8 | 8 |
| Total | 9 | 14 | 23 |

Logistic regression models how a {0,1} dichotomy like *any* depends on one or more $X$ variables.  The syntax of **logit** resembles that of **regress**, listing dependent variable first:

```
. logit any date
```

```
Iteration 0:   Log Likelihood =-15.394543
Iteration 1:   Log Likelihood = -13.01923
Iteration 2:   Log Likelihood =-12.991146
Iteration 3:   Log Likelihood =-12.991096
```

```
Logit Estimates                                  Number of obs =      23
                                                 chi2(1)       =    4.81
                                                 Prob > chi2   = 0.0283
Log Likelihood = -12.991096                      Pseudo R2     = 0.1561

-----------------------------------------------------------------------------
    any |    Coef.   Std. Err.       t     P>|t|     [95% Conf. Interval]
--------+--------------------------------------------------------------------
   date |   .0020907   .0010703    1.953   0.064     -.000135    .0043165
  _cons |  -18.13116   9.517217   -1.905   0.071     -37.9233    1.660974
-----------------------------------------------------------------------------
```

logit's iterative estimation procedure seeks to maximize the logarithm of the likelihood function, as shown at the output's top. The iteration 0 log likelihood describes the fit of a model including only the constant. The last log likelihood describes the fit of the final model:

$$L = -18.13116 + .0020907 \times date \qquad [11.1]$$

where $L$ represents the predicted logit, or log odds, of any distress incidents:

$$L = \ln[P(any=1) / P(any=0)] \qquad [11.2]$$

An overall $\chi^2$ test at upper right evaluates the null hypothesis that all coefficients in the model, except the constant, equal zero:

$$\chi^2 = -2(\ln \mathcal{L}_i - \ln \mathcal{L}_f) \qquad [11.3]$$

where $\ln \mathcal{L}_i$ is the initial or iteration 0 (model with constant only) log likelihood, and $\ln \mathcal{L}_f$ is the final iteration's log likelihood. Here:

$$\chi^2 = -2[-15.394543 - (-12.991096)]$$
$$= 4.81$$

The probability of a greater $\chi^2$, with 1 degree of freedom (the difference in complexity between initial and final models), is low enough (.0283) to reject the null hypothesis in this example.

Less accurate, though convenient, tests are provided by the asymptotic $t$ statistics displayed with logit results. With one $X$ variable, that variable's $t$ statistic and the overall $\chi^2$ statistic test equivalent hypotheses, analogous to the usual $t$ and $F$ statistics in bivariate OLS regression. Unlike their OLS counterparts, the logit $t$ approximation and $\chi^2$ tests sometimes disagree (they do here). The $\chi^2$ test has more general validity.

Like Stata's other maximum-likelihood estimation procedures, logit displays a pseudo $R^2$ with its output:

$$\text{pseudo } R^2 = 1 - \ln \mathcal{L}_f / \ln \mathcal{L}_i \qquad [11.4]$$

For this example:

$$\text{pseudo } R^2 = 1 - (-12.991096) / (-15.394543)$$
$$= .1561$$

Unfortunately, pseudo $R^2$ statistics lack the straightforward explained-variance interpretation of true $R^2$ in OLS regression.

After `logit`, the `predict` command (with no options) obtains predicted probabilities:

$$Phat = 1 / (1 + e^{-L})$$    [11.5]

Graphed against *date*, these probabilities follow an S-shaped logistic curve (Figure 11.1):

```
. predict Phat
. label variable Phat "predicted P(distress>=1)"
. graph Phat date, connect(s)
```

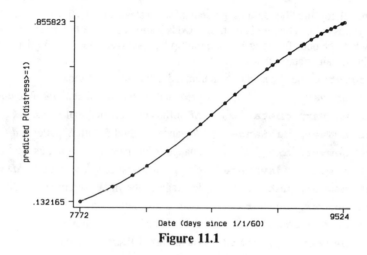

**Figure 11.1**

The coefficient given by `logit` ( .0020907) describes *date*'s effect on the logit or log odds of any thermal distress incidents. Each additional day increased the predicted log odds of thermal distress incidents by .0020907. Equivalently, we could say that each additional day multiplied predicted odds of thermal distress by $e^{.0020907} = 1.0020929$; each 100 days therefore multiplied the odds by $(e^{.0020907})^{100} \approx 1.23$. ($e \approx 2.71828$, the base number for natural logarithms.) Stata can make these calculations utilizing the `_b[]` coefficients stored after any estimation:

```
. display exp(_b[date])
1.0020929
. display exp(_b[date])^100
1.2325359
```

An easier approach, though, employs Stata's `logistic` command. `logistic` performs exactly the same estimation as `logit`, but it displays the results as odds ratios rather than coefficients, and provides several useful diagnostic tools. (We could also get the odds ratios, but not the diagnostic tools, by using the `or` option: `logit any date, or` .)

## Using Logistic Regression

Here is the same regression seen earlier, but this time using `logistic` instead of `logit`:

```
. logistic any date
```

```
Logit Estimates                                  Number of obs =      23
                                                   chi2(1)     =    4.81
                                                   Prob > chi2 = 0.0283
Log Likelihood = -12.991096                        Pseudo R2    = 0.1561


----------------------------------------------------------------------
    any | Odds Ratio   Std. Err.      t     P>|t|    [95% Conf. Interval]
--------+-------------------------------------------------------------
   date |  1.002093    .0010725    1.953   0.064     .999865    1.004326
----------------------------------------------------------------------
```

Note the identical log likelihoods and $\chi^2$ statistics. Instead of coefficients (*b*), `logistic` displays odds ratios ($e^b$). The numbers in the "Odds Ratio" column of `logistic`'s output are amounts by which the odds favoring *Y*=1 are multiplied, per 1-unit increase in that *X* variable (if other *X* variables' values stay the same).

After `logistic`, the `lpredict` command offers these options:

| | |
|---|---|
| . `lpredict` *newvar* | predicted probability that *Y* variable equals 1 |
| . `lpredict` *newvar*, `dbeta` | $\Delta B$ influence statistic, analogous to Cook's *D* |
| . `lpredict` *newvar*, `deviance` | deviance residual for *j*th *X* pattern, $d_j$ |
| . `lpredict` *newvar*, `dx2` | change in Pearson $\chi^2$, written as $\Delta\chi^2$ or $\Delta\chi^2_P$ |
| . `lpredict` *newvar*, `ddeviance` | change in deviance $\chi^2$, written as $\Delta D$ or $\Delta\chi^2_D$ |
| . `lpredict` *newvar*, `hat` | leverage of the *j*th *X* pattern, $h_j$ |
| . `lpredict` *newvar*, `number` | assigns numbers to the *X* patterns, $j = 1,2,3, ... J$ |
| . `lpredict` *newvar*, `resid` | Pearson residual for *j*th *X* pattern, $r_j$ |
| . `lpredict` *newvar*, `rstandard` | standardized Pearson residual |

Influence statistics calculated by the `dbeta`, `dx2`, `ddeviance`, and `hat` options reflect not individual-case influence, but the consequences of dropping all cases with that same *X*-pattern. See Hosmer and Lemeshow (1989) for details. A later section of this chapter shows these statistics in use.

Does booster joint temperature also affect the probability of any distress incidents? We could investigate by including *temp* as a second predictor variable:

. `logistic` *any date temp*

```
Logit Estimates                                  Number of obs =      23
                                                   chi2(2)     =    8.09
                                                   Prob > chi2 = 0.0175
Log Likelihood = -11.350748                        Pseudo R2    = 0.2627


----------------------------------------------------------------------
    any | Odds Ratio   Std. Err.      t     P>|t|    [95% Conf. Interval]
--------+-------------------------------------------------------------
   date |  1.00297     .0013675    2.175   0.042    1.000121    1.005826
   temp |  .8408309    .0987887   -1.476   0.156     .6580705    1.074348
----------------------------------------------------------------------
```

Each 1-degree increase in joint temperature multiplies the odds of booster joint damage by .84 (that is, each 1-degree warming reduces the odds of damage by about 16%). Although this effect seems strong enough to cause concern, the asymptotic *t* test says that it is not statistically significant ($t = -1.476$, $P = .156$).

We can perform a better test of *temp*'s effect with a likelihood-ratio $\chi^2$. The `lrtest` command compares nested models estimated by maximum likelihood. First, estimate a "complete" model containing all variables of interest, as done above with the `logistic` *any date temp* command. Next, type the command:

```
. lrtest, saving(0)
```

Now estimate a reduced model, including only a subset of the *X* variables from the complete model. (Such reduced models are said to be "nested.") Finally, type `lrtest` again. For example (using the `quietly` prefix, since we already saw this output once):

```
. quietly logistic any date
. lrtest
Logistic:   likelihood-ratio test          chi2(1)     =      3.28
                                            Prob > chi2 =    0.0701
```

This second `lrtest` command tests the recent (presumably nested) model against the model previously saved by `lrtest, saving(0)`. It employs a general test statistic for nested maximum-likelihood models:

$$\chi^2 = -2(\ln \mathcal{L}_1 - \ln \mathcal{L}_0) \qquad [11.6]$$

where $\ln \mathcal{L}_0$ is the log likelihood for the first model (with all *X* variables), and $\ln \mathcal{L}_1$ the log likelihood for the second model (with a subset of those *X* variables). Compare the resulting test statistic to a $\chi^2$ distribution with degrees of freedom equal to the difference in complexity (number of *X* variables dropped) between models 0 and 1. Type `help lrtest` for more about this command, which works with any of Stata's maximum-likelihood estimation procedures (`logit`, `probit`, `cox`, etc.). The overall $\chi^2$ statistic routinely given by `logit` or `logistic` output (equation [11.3]) is a special case of [11.6].

The `lrtest` example above performed this calculation:

$$\chi^2 = -2[-12.991096 - (-11.350748)]$$
$$= 3.28$$

with 1 degree of freedom, yielding *P* = .0701; *temp*'s effect is significant at $\alpha = .10$. Given the small sample and fatal consequences of a Type II error, $\alpha = .10$ seems a more prudent cutoff than the usual $\alpha = .05$.

## Conditional Effect Plots

Conditional effect plots help in understanding what a logit or logistic model implies for probabilities. For example, `summarize date, detail` tells us that the 25th percentile of *date* equals 8569. To find the predicted probability of any distress incidents, as a function of *temp*, with *date* fixed at its 25th percentile:

```
. quietly logit any date temp
. generate L1 = _b[_cons]+_b[date]*8569+_b[temp]*temp
. generate Phat1 = 1/(1+exp(-L1))
. label variable Phat1 "P(distress>=1|date=8569)"
```

*L1* is the predicted logit, and *Phat1* equals the corresponding predicted probability, calculated according to equation [11.5]. Similar steps find the predicted probability of any distress with date fixed at its 75th percentile (9341):

```
. generate L2 = _b[_cons]+_b[date]*9341+_b[temp]*temp
. generate Phat2 = 1/(1+exp(-L2))
. label variable Phat2 "P(distress>=1|date=9341)"
```

We can now graph the relation between *temp* and the probability of any distress, for the two levels of *date* (Figure 11.2):

```
. graph Phat1 Phat2 temp, connect(ss) ylabel xlabel
        11(Probability of distress>=1)
```

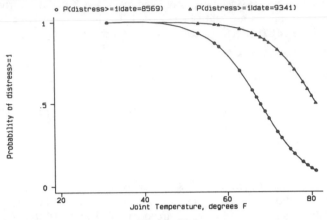

**Figure 11.2**

Among earlier flights (*date* = 8569, left curve), probability of thermal distress goes from very low, at around 80° F, to near 1, below 50° F. Among later flights (*date* = 9341, right curve), however, the probability of any distress exceeds .5 even in warm weather, and climbs toward 1 on flights below 70° F. Note that *Challenger*'s launch temperature, 31° F, places it at top left in Figure 11.2; the analysis predicts almost certain booster joint damage.

## Diagnostic Statistics and Graphs

The influence and diagnostic statistics calculated by `lpredict` (after `logistic`) refer not to individual cases, as do the regression diagnostics of Chapter 8. Rather, the `lpredict` diagnostics refer to *X* patterns: specific combinations of *X* values, that may be shared by any number of cases. With the space shuttle data, however, each *X* pattern is unique—no two flights share the same combination of *date* and *temp* (naturally, since no two were launched the same day).

Hosmer and Lemeshow (1989) suggest plots that help in reading these diagnostics. Before generating diagnostic statistics, we might `drop` unneeded variables from our data:

```
. drop Phat L1-Phat2
```

Such housecleaning is often necessary using Student Stata, which can hold only 25 variables. Next we calculate and label some diagnostic variables. Since `lpredict` uses results from the most recent `logistic` estimation, we begin by quietly repeating this command, to be sure it is what we think:

```
. quietly logistic any date temp
. lpredict Phat
. label variable Phat "Predicted Probability"
. lpredict dX2, dx2
. label variable dX2 "Change in Pearson Chi-Square"
. lpredict dB, dbeta
. label variable dB "Influence"
. lpredict dD, ddeviance
. label variable dD "Change in Deviance"
```

To graph change in Pearson $\chi^2$ versus predicted probability of distress (Figure 11.3), type:

. graph *dX2 Phat*, ylabel xlabel

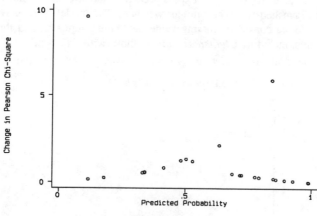

**Figure 11.3**

Two poorly fit $X$ patterns, at upper right and left, stand out. We can identify these two flights (STS-2 and STS 51-A) by drawing the graph with flight numbers as plotting symbols. **symbol([*flight*]) psize(150)** accomplishes this, showing the flight numbers at 150% of the usual plotting-symbol size (Figure 11.4):

. graph *dX2 Phat*, ylabel xlabel symbol([*flight*]) psize(150)

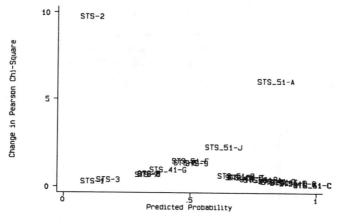

**Figure 11.4**

. list *flight any date temp dX2 Phat if dX2 > 5*

|     | flight   | any | date | temp | dX2      | Phat     |
|-----|----------|-----|------|------|----------|----------|
| 22. | STS_51-A | 0   | 9078 | 67   | 5.899742 | .8400974 |
| 23. | STS-2    | 1   | 7986 | 70   | 9.630337 | .1091805 |
| 24. | STS-4    | .   | 8213 | 80   | .        | .        |
| 25. | STS_51-L | .   | 9524 | 31   | .        | .        |

Flight STS 51-A experienced no thermal distress, despite late launch date and cool temperature (see Figure 11.2). The model predicts a .84 probability of distress for this flight. All points along the up-to-right curve in Figure 11.4 have *any* = 0, meaning no thermal distress. Atop the up-to-left (*any* = 1) curve, flight STS-2 experienced thermal distress despite being one of the earliest flights, and launched in slightly milder weather. The model predicts only a .109 probability of distress. (Stata considers missing values as "high" numbers, and thus `listed` two missing-values flights, including *Challenger*, among those with *dX2* > 5.)

Similar findings result from plotting *dD* ( Δ*D* ) versus predicted probability (Figure 11.5):

```
. graph dD Phat, ylabel xlabel symbol([flight])
```

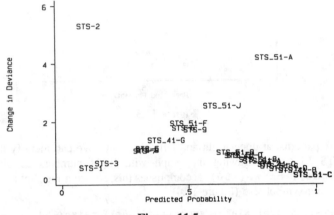

**Figure 11.5**

Again, flights STS-2 (top left) and STS 51-A (top right) stand out as poorly fit.

*dB* measures an *X* pattern's influence in logistic regression, as Cook's *D* measures case influence in OLS. For a logistic-regression analogue to the OLS diagnostic plot in Figure 8.8 (page 106), we can make the plotting symbols proportional to influence (Figure 11.6):

```
. graph dD Phat [iweight=dB], ylabel xlabel
```

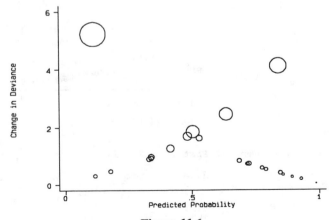

**Figure 11.6**

Figure 11.6 reveals that the two worst-fit cases are also the most influential.

Cases poorly fit <u>and</u> influential deserve special attention, because they both contradict the main pattern of the data and pull model estimates in their contrary direction. Of course, simply removing such outliers allows a "better fit" with the remaining data—but this is circular reasoning. A more thoughtful reaction would be to investigate what makes the outliers unusual. Why did shuttle flight STS-2, but not STS 51-A, experience booster joint damage? Seeking an answer might lead investigators to previously overlooked variables, or to respecifying the model.

## Logistic Regression with Ordered-Category *Y*

`logit` and `logistic` estimate models with two-category {0,1} *Y* variables. We need other methods to estimate models in which *Y* takes on more than two categories:

`ologit` ordered logistic regression, where *Y* is an ordinal (ordered-category) variable. The numerical values representing the categories do not matter, except that higher numbers mean "more." For example, the *Y* categories might be {1="poor", 2="fair", 3="excellent"}.

`mlogit` multinomial logistic regression, where *Y* has multiple but unordered categories like {1="Democrat", 2="Republican", 3="undeclared"}.

If *Y* is {0,1}, `logit`, `ologit`, and `mlogit` produce essentially the same estimates.

We earlier simplified the three-category ordinal variable *distress* into a two-category {0,1} dichotomy, *any*. `logit` and `logistic` require {0,1} dependent variables. `ologit`, on the other hand, can handle ordinal variables like *distress* that have more than two categories. The numerical codes representing these categories do not matter, so long as higher numerical values mean "more" of whatever is being measured. Recall that *distress* has categories 0="none", 1="1 or 2", and 2="3 plus" incidents of booster-joint distress.

Ordered-logit regression indicates that *date* and *temp* both affect *distress*, with the same signs (positive for *date*, negative for *temp*) seen in our earlier analyses:

```
. ologit distress date temp

Iteration 0:   Log Likelihood =-24.955257
Iteration 1:   Log Likelihood =-19.093131
Iteration 2:   Log Likelihood =-18.805891
Iteration 3:   Log Likelihood =-18.797073
Iteration 4:   Log Likelihood = -18.79706

Ordered Logit Estimates                    Number of obs =      23
                                           chi2(2)       =   12.32
                                           Prob > chi2   =  0.0021
Log Likelihood =  -18.79706                Pseudo R2     =  0.2468

------------------------------------------------------------------------
distress |     Coef.   Std. Err.      t      P>|t|    [95% Conf. Interval]
---------+--------------------------------------------------------------
    date |   .003286   .0012662     2.595    0.017    .0006528    .0059192
    temp | -.1733752   .0834473    -2.078    0.050   -.3469135     .000163
---------+--------------------------------------------------------------
   _cut1 |  16.42813   9.554813          (Ancillary parameters)
   _cut2 |  18.12227   9.722293
------------------------------------------------------------------------
```

Likelihood-ratio tests are more accurate than the asymptotic *t* tests shown. First, estimate and have `lrtest` save results from the "full" model including both predictors:

```
. quietly ologit distress date temp
. lrtest, saving(0)
```

Next, estimate a simpler model without *temp*:

```
. quietly ologit distress date
. lrtest
Ologit:  likelihood-ratio test                    chi2(1)     =      6.12
                                                   Prob > chi2 =    0.0133
```

The likelihood-ratio test indicates that *temp*'s effect is significant. Similar steps find that *date* also has a significant effect:

```
. quietly ologit distress temp
. lrtest
Ologit:  likelihood-ratio test                    chi2(1)     =     10.33
                                                   Prob > chi2 =    0.0013
```

The ordered-logit model estimates a score, *S*, as a linear function of *date* and *temp*:

$$S = .003286 \times date - .1733752 \times temp$$

Predicted probabilities depend on the value of *S* (plus a logistically distributed disturbance *u*) relative to the estimated cut points:

$P(distress="none")$ $= P(S+u \leq \_cut1)$ $= P(S+u \leq 16.42813)$
$P(distress="1 \text{ or } 2")$ $= P(\_cut1 < S+u \leq \_cut2)$ $= P(16.42813 < S+u \leq 18.12227)$
$P(distress="3 \text{ plus}")$ $= P(\_cut2 < S+u)$ $= P(18.12227 < S+u)$

After **ologit**, the command **ologitp** calculates predicted probabilities for each category of the dependent variable. We supply **ologitp** with names for these probabilities. For example: *none* could denote the probability of no distress incidents (first category of *distress*); *onetwo* the probability of 1–2 incidents (second category of *distress*); and *threeplu* the probability of 3 or more incidents (third and last category of *distress*):

```
. quietly ologit distress date temp
. ologitp none onetwo threeplu
```

This creates three new variables:

```
. describe

Contains data from c:\stustata\shuttle.dta
  Obs:      25 (max=  2620)              Space Shuttle data
  Vars:     11 (max=    99)
 Width:     22 (max=   200)
   1. flight      byte   %8.0g   flbl   Flight
   2. month       byte   %8.0g          Month of Launch
   3. day         byte   %8.0g          Day of Launch
   4. year        int    %8.0g          Year of Launch
   5. distress    byte   %8.0g   dlbl   Thermal Distress Incidents
   6. temp        byte   %8.0g          Joint Temperature, degrees F
   7. date        int    %8.0g          Date (days since 1/1/60)
   8. any         byte   %8.0g          Any Thermal Distress
   9. none        float  %9.0g          Pr(xb+u<_cut1)
  10. onetwo      float  %9.0g          Pr(_cut1<xb+u<_cut2)
  11. threeplu    float  %9.0g          Pr(_cut2<xb+u)
Sorted by:
Note:  Data has changed since last save
```

Predicted probabilities for *Challenger*'s last flight, the 25th in these data, are unsettling:

```
. list if flight==25
```

```
Observation 25

      flight    STS_51-L      month              1      day          28
       year        1986      distress            .      temp         31
       date        9524         any              .      none .0000753686
     onetwo  .0003346481    threeplu  .9995899796
```

Our model, based on analysis of 23 pre-*Challenger* shuttle flights, predicts little chance (*P* = .000075) of *Challenger* experiencing no booster joint damage; a scarcely greater likelihood of one or two incidents (*P* = .0003); but virtual certainty (*P* = .9996) of three or more damage incidents.

See the *Stata Reference Manual* or Greene (1990) for more about ordered logistic regression.

## Multinomial Logistic Regression

When the dependent variable's categories have no natural ordering, we resort to multinomial logistic regression (also called polytomous logistic regression). Dataset *dover.dta* contains information from an environmental-issues survey of registered voters in Dover, New Hampshire:

```
. use c:\stustata\dover, clear
(5/85 Dover water survey)
. describe

Contains data from c:\stustata\dover.dta
  Obs:    150 (max=  2620)                  5/85 Dover water survey
  Vars:     9 (max=    99)
 Width:     9 (max=   200)
    1. age      byte   %8.0g              years of age
    2. educ     byte   %8.0g              respondent's education in years
    3. sex      byte   %8.0g    sexrlbl   respondent's sex (female)
    4. kids     byte   %8.0g    anyklbl   have kids <18 in household?
    5. party    byte   %8.0g    partylbl  political party registration
    6. zoning   byte   %8.0g    zonlbl    pass aquifer zoning
    7. superf   byte   %8.0g    superlbl  for Superfund cleanup Tolend Rd
    8. mapping  byte   %8.0g    maplbl    detailed mapping of groundwater
    9. cleanup  byte   %8.0g    hiplbl    participated in household picku
Sorted by:
```

The variable *party* records respondents' political affiliation:

```
. tabulate party
    political|
       party|
registration|     Freq.      Percent       Cum.
------------+-----------------------------------
    Democrat |       45        30.00       30.00
    Republic |       62        41.33       71.33
    undeclar |       43        28.67      100.00
------------+-----------------------------------
      Total  |      150       100.00
```

*party* is actually a labeled numeric variable, as revealed by typing:

```
. tabulate party, nolabel
    political|
       party|
registration|     Freq.      Percent       Cum.
------------+-----------------------------------
           1 |       45        30.00       30.00
           2 |       62        41.33       71.33
           3 |       43        28.67      100.00
------------+-----------------------------------
      Total  |      150       100.00
```

The coding {1="Democrat", 2="Republican", 3="undeclared"} has no numerical meaning. We could equally well have used {1="Republican", 5="undeclared", 29="Democrat"}, or any other scheme.

Is there a "gender gap" among Dover voters? An ordinary (Pearson) $\chi^2$ test finds no association between *party* and *sex*:

```
. tabulate party sex, chi2

political| respondent's sex (female)
    party|
registratio|
        n|    male    female |    Total
-----------+--------------------+----------
 Democrat |     18       27 |       45
 Republic |     33       29 |       62
 undeclar |     18       25 |       43
-----------+--------------------+----------
    Total|     69       81 |      150

        Pearson chi2(2) =   2.2520   Pr = 0.324
```

Multinomial logistic regression can replicate this simple analysis, obtaining a similar $\chi^2$ value from its likelihood-ratio test:

```
. mlogit party sex, rrr nolog

Multinomial regression                        Number of obs =      150
                                              chi2(2)       =     2.25
                                              Prob > chi2   = 0.3240
Log Likelihood = -161.55453                   Pseudo R2     = 0.0069

------------------------------------------------------------------------
    party |     RRR    Std. Err.      t    P>|t|    [95% Conf. Interval]
----------+-------------------------------------------------------------
Democrat |
     sex |  1.706897  .6771427    1.348   0.180    .7793002   3.738605
----------+-------------------------------------------------------------
undeclar |
     sex |  1.58046   .632857     1.143   0.255    .7162943   3.487188
------------------------------------------------------------------------
(Outcome party==Republic is the comparison group)
```

Note the output's bottom line: *party*="Republican" is the comparison group or base category. Unless we tell it otherwise, **mlogit** automatically chooses the most frequent category (in this instance, Republicans) as a base. The **rrr** option instructs **mlogit** to show relative risk ratios, which resemble the odds ratios given by **logistic**. Another option, **nolog**, suppresses printing the iteration log.

Using the **tabulate** output at the top of this page, we can calculate that *among males* the odds favoring Democrat over Republican are:

$P$(Democrat) / $P$(Republican)    = (18/69) / (33/69)
                                   = .5454545

*Among females* the odds favoring Democrat over Republican are:

$P$(Democrat) / $P$(Republican)    = (27/81) / (29/81)
                                   = .9310344

Thus the odds favoring Democrat over Republican are

.9310344 / .5454545 = 1.706897

times higher for females (*sex*=1) than for males (*sex*=0). This multiplier, a ratio of two odds, equals the relative risk ratio (1.706897) displayed by **mlogit**.

In general, the relative risk ratio for category $j$ of $Y$, and predictor $X_k$, equals the amount by which predicted odds favoring $Y = j$ (compared with $Y$ = base) are multiplied, per 1-unit increase in $X_k$, other things being equal. In other words, the relative risk ratio $rrr_{jk}$ is a multiplier such that, if all $X$ variables except $X_k$ stay the same:

$$rrr_{jk} \times \frac{P(Y = j \mid X_k)}{P(Y = \text{base} \mid X_k)} = \frac{P(Y = j \mid X_k+1)}{P(Y = \text{base} \mid X_k+1)} \qquad [11.7]$$

We can override the default base category by specifying **base(#)** as an option. Since "undeclared" is category 3 of *party*, **base(3)** makes "undeclared" the **mlogit** base category.

. **mlogit** *party sex*, **rrr base(3) nolog**

```
Multinomial regression                          Number of obs =      150
                                                   chi2(2)    =     2.25
                                                   Prob > chi2 = 0.3240
Log Likelihood = -161.55453                        Pseudo R2   = 0.0069
```

| party | RRR | Std. Err. | t | P>\|t\| | [95% Conf. Interval] | |
|-------|-----|-----------|---|---------|----------------------|--|
| Democrat | | | | | | |
| sex | 1.08 | .4684613 | 0.177 | 0.859 | .4582687 | 2.545232 |
| Republic | | | | | | |
| sex | .6327273 | .2533607 | -1.143 | 0.255 | .2867637 | 1.396076 |

(Outcome party==undeclar is the comparison group)

Changing the base category changes relative risk ratios, but has no effect on the model fit, predictions, or $\chi^2$ statistic.

Both **tabulate** and **mlogit** (with any base category) agree that gender is not a good predictor of political party affiliation. Age and education exhibit stronger effects:

. **mlogit** *party age educ*, **rrr base(3) nolog**

```
Multinomial regression                          Number of obs =      150
                                                   chi2(4)    =    34.58
                                                   Prob > chi2 = 0.0000
Log Likelihood = -145.39136                        Pseudo R2   = 0.1063
```

| party | RRR | Std. Err. | t | P>\|t\| | [95% Conf. Interval] | |
|-------|-----|-----------|---|---------|----------------------|--|
| Democrat | | | | | | |
| age | 1.077405 | .0197447 | 4.068 | 0.000 | 1.039076 | 1.117147 |
| educ | 1.266085 | .1322149 | 2.259 | 0.025 | 1.029959 | 1.556344 |
| Republic | | | | | | |
| age | 1.088564 | .0193815 | 4.766 | 0.000 | 1.050921 | 1.127555 |
| educ | 1.248008 | .1252581 | 2.207 | 0.029 | 1.023437 | 1.521856 |

(Outcome party==undeclar is the comparison group)

The approximate $t$ tests suggest that both *age* and *educ* have significant effects, but a likelihood-ratio test is more convincing. To test the effect of education, compare the full model just estimated with a reduced model, lacking *educ* but otherwise the same:

```
. lrtest, saving(0)
. quietly mlogit party age
. lrtest
Mlogit:  likelihood-ratio test                   chi2(2)    =      6.56
                                                  Prob > chi2 =    0.0376
```

We can reject the null hypothesis that *educ* has no effect. (The **base(#) rrr nolog** options do not matter here, because they do not change likelihood ratios.) We can also test the effect of *age*, by comparing the full model to a reduced model without *age*:

```
. quietly mlogit party educ
. lrtest
Mlogit:  likelihood-ratio test                   chi2(2)    =     32.89
                                                  Prob > chi2 =    0.0000
```

Both *age* and *educ* significantly predict *party*. What else can we say? To interpret these effects, recall that "undeclared" is now the base category. The relative risk ratios tell us that:

Odds of "Democrat" rather than "undeclared" are multiplied by 1.077 (increase about 8%) with each 1-year increase in age, controlling for education.

Odds of "Democrat" rather than "undeclared" are multiplied by 1.266 (increase about 27%) with each 1-year increase in education, controlling for age.

Odds of "Republican" rather than "undeclared" are multiplied by 1.089 (increase about 9%) with each 1-year increase in age, controlling for education.

Odds of "Republican" rather than "undeclared" are multiplied by 1.248 (increase about 25%) with each 1-year increase in education, controlling for age.

Thus being older and/or better educated increases the odds of registering for one of the political parties, rather than "undeclared."

Thus *age* and *educ* predict "Democrat" versus "undeclared," and "Republican" versus "undeclared." But neither *age* nor *educ* has much effect on the odds of "Republican" over "Democrat," as we see by changing the base category:

```
. mlogit party age educ, rrr base(1) nolog
```

```
Multinomial regression                      Number of obs =       150
                                               chi2(4)    =     34.58
                                               Prob > chi2 =    0.0000
Log Likelihood = -145.39136                    Pseudo R2   =    0.1063

-----------------------------------------------------------------------
  party |      RRR    Std. Err.        t     P>|t|    [95% Conf. Interval]
--------+--------------------------------------------------------------
Republic |
    age | 1.010358   .0144412      0.721    0.472     .9822129   1.039309
   educ | .9857225   .085046      -0.167    0.868      .831175   1.169006
--------+--------------------------------------------------------------
undeclar |
    age | .9281564   .0170096     -4.068    0.000     .8951373   .9623934
   educ | .7898367   .0824812     -2.259    0.025     .6425317   .9709124
-----------------------------------------------------------------------
(Outcome party==Democrat is the comparison group)
```

Nor do *age* or *educ* affect the odds of "Democrat" over "Republican":

```
. mlogit party age educ, rrr base(2) nolog
```

Multinomial regression

Log Likelihood = -145.39136

```
Number of obs =      150
chi2(4)       =    34.58
Prob > chi2   = 0.0000
Pseudo R2     = 0.1063
```

| party | RRR | Std. Err. | t | P>\|t\| | [95% Conf. Interval] |
|---|---|---|---|---|---|
| Democrat |  |  |  |  |  |
| age | .9897486 | .0141467 | -0.721 | 0.472 | .962178    1.018109 |
| educ | 1.014484 | .0875275 | 0.167 | 0.868 | .8554273   1.203116 |
| undeclar |  |  |  |  |  |
| age | .9186415 | .0163561 | -4.766 | 0.000 | .8868746   .9515461 |
| educ | .8012769 | .0804213 | -2.207 | 0.029 | .6570925   .9770994 |

(Outcome party==Republic is the comparison group)

**predict** can calculate predicted probabilities from **mlogit**. The **outcome(#)** option specifies for which *Y* category we want probabilities. For example, to get predicted probabilities that *party*="Republican" (category 2):

```
. quietly mlogit party age educ
. predict PRepub, outcome(2)
. label variable PRepub "predicted P(party=Republican)"
```

Predicted probabilities of registering Republican (*PRepub*) tend to be higher for actual Republicans than they are for undeclared voters. For Republicans and Democrats, though, the predicted probabilities of registering Republican are almost the same—because age and education did not much discriminate between Dover's voters from these two parties:

```
. tabulate party, summ(PRepub)
```

| political party registration | Summary of predicted P(party-Republican) Mean | Std. Dev. | Freq. |
|---|---|---|---|
| Democrat | .44265034 | .12437918 | 45 |
| Republic | .46324686 | .12532096 | 62 |
| undeclar | .3106844 | .13358532 | 43 |
| Total | .41333333 | .14270703 | 150 |

The *Stata Reference Manual* provides more details about multinomial logit. Introductions to this topic also appear in Aldrich and Nelson (1984), Greene (1990), and Hosmer and Lemeshow (1989).

## Also Type help

| | |
|---|---|
| clogit | conditional logistic regression |
| cnreg | censored-normal regression |
| constrain | define, list or drop linear constraints (e.g., with **mlogit**) |
| cox | proportional hazards model |
| dates | convert month-day-year to elapsed time, and other date functions |
| glogit | logistic regression from grouped data |
| linktest | specification link test of single-equation model |
| logistic | logistic regression, giving odds ratios and diagnostics |

| `logit` | logistic regression, giving estimated coefficients |
| `loglin` | loglinear modeling (supplied with Student Stata; see Judson 1992) |
| `lrtest` | likelihood-ratio $\chi^2$ test of nested maximum-likelihood models |
| `mlogit` | multinomial logistic regression, with polytomous $Y$ variable |
| `ologit` | logistic regression with ordered-category $Y$ variable |
| `oprobit` | probit regression with ordered-category $Y$ variable |
| `poisson` | poisson regression, assuming Poisson-distributed $Y$ variable |
| `predict` | predictions, some diagnostic statistics |
| `probit` | probit regression, with dichotomous $Y$ variable |
| `test` | tests of user-specified hypotheses |
| `tobit` | tobit regression |

# 12
# Principal Components and Factor Analysis

Principal components and factor analysis provide methods for simplification—combining many correlated variables into a smaller number of underlying dimensions. Along the way to achieving simplification, the analyst must choose from a daunting variety of options. If the data really do reflect clear-cut underlying dimensions, different options may nonetheless converge on similar results. In the absence of clear underlying dimensions, however, different options often lead to divergent results. Experimenting with these options can tell us how stable a particular finding is, or how much it depends on arbitrary choices about the specific analytical technique.

Stata accomplishes principal components and factor analysis with four basic commands:

**factor**     extracts principal components or several type of factors from a correlation or covariance matrix.

**greigen**    constructs a scree graph, or plot of the eigenvalues, from the recent **factor**.

**rotate**     performs orthogonal (uncorrelated factors) or oblique (correlated factors) rotation, after **factor**.

**score**      generates factor scores (composite variables) after **rotate** and/or **factor**.

The composite variables generated by **score** can subsequently be saved, listed, graphed, or analyzed like any other Stata variable.

## Principal Components

To illustrate basic principal components/factor analysis commands, we return to the planetary dataset introduced in Chapter 2. This version of the dataset includes a variable measuring density and also a set of the variables in natural log form:

```
. use c:\stustata\planets2, clear
(Solar system data)
. describe

Contains data from c:\stustata\planets2.dta
  Obs:      9 (max=  2620)                Solar system data
  Vars:    12 (max=    99)
Width:     53 (max=   200)
   1. planet      str7    %9s              Planet
   2. dsun        float   %9.0g            Mean dist. sun, km*10^6
   3. radius      float   %9.0g            Equatorial radius in km
   4. rings       byte    %8.0g    ringlbl Has rings?
   5. moons       byte    %8.0g            Number of known moons
   6. mass        float   %9.0g            Mass in kilograms
   7. density     float   %9.0g            Mean density, g/cm^3
   8. logdsun     float   %9.0g            natural log dsun
   9. lograd      float   %9.0g            natural log radius
  10. logmoons    float   %9.0g            natural log (moons + 1)
  11. logmass     float   %9.0g            natural log mass
  12. logdense    float   %9.0g            natural log density
Sorted by:  dsun
```

To extract initial factors or principal components, use the command `factor` followed by a variable list (variables in any order) and one of these options:

| | |
|---|---|
| `pc` | principal components |
| `pcf` | principal components factoring |
| `pf` | principal factoring (default) |
| `ipf` | principal factoring with iterated communalities |
| `ml` | maximum-likelihood factoring |

For example, to obtain principal components factors:

```
. factor rings logdsun-logdense, pcf
(obs=9)
```

(principal component factors; 2 factors retained)

| Factor | Eigenvalue | Difference | Proportion | Cumulative |
|---|---|---|---|---|
| 1 | 4.62365 | 3.45469 | 0.7706 | 0.7706 |
| 2 | 1.16896 | 1.05664 | 0.1948 | 0.9654 |
| 3 | 0.11232 | 0.05395 | 0.0187 | 0.9842 |
| 4 | 0.05837 | 0.02174 | 0.0097 | 0.9939 |
| 5 | 0.03663 | 0.03657 | 0.0061 | 1.0000 |
| 6 | 0.00006 | . | 0.0000 | 1.0000 |

Factor Loadings

| Variable | 1 | 2 | Uniqueness |
|---|---|---|---|
| rings | 0.97917 | 0.07720 | 0.03526 |
| logdsun | 0.67105 | -0.71093 | 0.04427 |
| lograd | 0.92287 | 0.37357 | 0.00875 |
| logmoons | 0.97647 | 0.00028 | 0.04651 |
| logmass | 0.83377 | 0.54463 | 0.00821 |
| logdense | -0.84511 | 0.47053 | 0.06439 |

Only the first two components have eigenvalues greater than 1, and these two components explain over 96% of the six variables' combined variance. The unimportant 3rd through 6th principal components might safely be disregarded in subsequent analysis.

Two `factor` options provide control over the number of factors extracted:

> `factors(#)`    where # specifies the number of factors

> `mineigen(#)`    where # specifies the minimum eigenvalue for retained factors

The principal components factoring (`pcf`) procedure automatically drops factors with eigenvalues below 1, so

```
. factor rings logdsun-logdense, pcf
```

is equivalent to

```
. factor rings logdsun-logdense, pcf mineigen(1)
```

In this example, we would also have obtained the same results by typing:

```
. factor rings logdsun-logdense, pcf factors(2)
```

To see a scree graph (plot of eigenvalues versus component or factor number) after any `factor`, simply type:

```
. greigen
```

`graph` options can be added to `greigen`, to draw a more presentable image (Figure 12.1):

```
. greigen, yline(1) ylabel(0,1,2,3,4,5) xlabel(1,2,3,4,5,6)
          b2(Component Number)
```

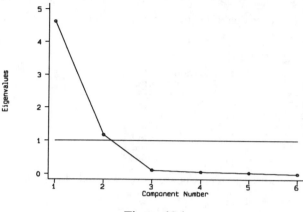

**Figure 12.1**

Figure 12.1 again emphasizes the unimportance of components 3–6.


# Rotation

Rotation aims to further simplify factor structure. After factoring, type **rotate** followed by either:

**varimax**      varimax orthogonal rotation, for uncorrelated factors or components (default).

**promax(#)**    promax oblique rotation, allowing correlated factors or components. Choose a number (promax power) $\leq 4$; the higher the number, the greater the degree of interfactor correlation. **promax(3)** is the default.

Two other **rotate** options are:

**factors(#)**    as it does with **factor**, this option specifies how many factors to retain.

**horst**        Horst modification to varimax and promax rotation.

Rotation (and factor scoring) can be performed following any factor analysis, whether it employed the **pcf**, **pf**, **ipf**, or **ml** options. In this section, we will follow through on our **pcf** example. For orthogonal (default) rotation of the first two components found in the planetary data:

```
. rotate
```

```
               (varimax rotation)
              Rotated Factor Loadings
   Variable |     1          2      Uniqueness
-----------+--------------------------------
      rings |  -0.74739    0.63729    0.03526
    logdsun |   0.02755    0.97723    0.04427
     lograd |  -0.91698    0.38781    0.00875
   logmoons |  -0.69112    0.68981    0.04651
    logmass |  -0.97481    0.20381    0.00821
   logdense |   0.26549   -0.93012    0.06439
```

This example accepts all the defaults: varimax rotation and the same number of factors retained in the last **factor**. We could have asked for the same rotation explicitly, by the command:

```
. rotate, varimax factors(2)
```

For oblique promax rotation of the most recent factoring:

```
. rotate, promax
                  (promax rotation)
                Rotated Factor Loadings
    Variable |      1          2        Uniqueness
    ---------+---------------------------------------
       rings |  -0.70990     0.41625      0.03526
     logdsun |   0.23272     1.06946      0.04427
      lograd |  -0.95231     0.08350      0.00875
    logmoons |  -0.63571     0.49386      0.04651
     logmass |  -1.05540    -0.13721      0.00821
    logdense |   0.10691    -0.91073      0.06439
```

By default, this example used a promax power of 3.  We could have asked for this explicitly:

```
. rotate, promax(3) factors(2)
```

**promax(4)** would permit further simplification of the loading matrix, at the cost of stronger interfactor correlations and hence less simplification.

After promax rotation, *rings*, *lograd*, *logmoons*, and *logmass* load most heavily (and all negatively) on factor 1.  This appears to be a "smallness/few satellites" dimension. *logdsun* and *logdense* load higher on factor 2, forming a "far out/low density" dimension.  The next section shows how to create new variables representing these dimensions.

## Factor Scores

Factor scores are linear composites, formed by standardizing each variable to zero mean and unit variance, then weighting with factor score coefficients and summing for each factor.  **score** performs these calculations automatically, using the most recent **rotate** or **factor** results:

```
. score f1 f2
              (based on rotated factors)
            Scoring Coefficients
    Variable |      1          2
    ---------+---------------------
       rings |  -0.21915     0.13667
     logdsun |   0.13871     0.46517
      lograd |  -0.32032    -0.01750
    logmoons |  -0.18953     0.17360
     logmass |  -0.36774    -0.11655
    logdense |  -0.01402    -0.37922
```

```
. label variable f1 "smallness/few satellites"
. label variable f2 "far out/low density"
```

We supply names for the new variables, unimaginatively called *f1* and *f2* above.

```
. list planet f1 f2
```

|    | planet  | f1        | f2        |
|----|---------|-----------|-----------|
| 1. | Mercury | .8522902  | -1.273661 |
| 2. | Venus   | .4416931  | -1.182095 |
| 3. | Earth   | .3271711  | -1.025886 |
| 4. | Mars    | .6580691  | -.62064   |
| 5. | Jupiter | -1.361802 | .4573379  |
| 6. | Saturn  | -1.155303 | .9708819  |
| 7. | Uranus  | -.7230982 | .9527438  |
| 8. | Neptune | -.6068392 | .8303058  |
| 9. | Pluto   | 1.567819  | .8910129  |

Being standardized variables, the new factor scores have means (approximately) equal to zero and standard deviations equal to one:

```
. summ f1 f2

Variable |    Obs        Mean   Std. Dev.        Min         Max
---------+------------------------------------------------------------
      f1 |      9    3.31e-09           1   -1.361802    1.567819
      f2 |      9    3.31e-09           1   -1.273661     .9708819
```

Thus the factor scores are measured in units of standard deviations from their means. Mercury, for example, is about .85 standard deviations above average on the smallness/few satellites dimension, and 1.27 standard deviations below average on far out/low density dimension.

Promax rotation permits correlations between factor scores:

```
. correlate f1 f2
(obs=9)
         |      f1        f2
---------+------------------
      f1 |  1.0000
      f2 | -0.4865    1.0000
```

Scores on factor 1 have a moderate negative correlation with scores on factor 2: smaller planets are less likely to be far out/low density.

If we employed varimax instead of promax rotation, we would get uncorrelated factor scores. Using the **quietly** prefix to suppress output:

```
. quietly factor rings logdsun lograd logmoons logmass logdense, pcf
. quietly rotate
. quietly score varimax1 varimax2
. correlate varimax1 varimax2
(obs=9)
         | varimax1 varimax2
---------+------------------
varimax1 |  1.0000
varimax2 |  0.0000    1.0000
```

Once created by **score**, factor scores can be treated like any other Stata variable—listed, analyzed, graphed, and so forth. Graphs of principal component factors sometimes help in identifying multivariate outliers or clusters of cases that stand separate from the rest. For example, Figure 12.2 reveals three distinct types of planets:

```
. graph f1 f2, noaxis yline(0) xline(0) ylabel xlabel symbol([planet])
      psize(130)
```

The **symbol([planet]) psize(130)** options cause values of *planet* (planet name) to be used as plotting symbols, at 130% of normal plotting-symbol size. The inner, rocky planets (high on "smallness/few satellites" factor 1; low on "far out/low density" factor 2) plot together at upper left. The outer gas giants have opposite characteristics, and plot together at lower right. Pluto, which physically resembles some outer-system moons, is unique among planets for being high on both "smallness/few satellites" and "far out/low density" dimensions.

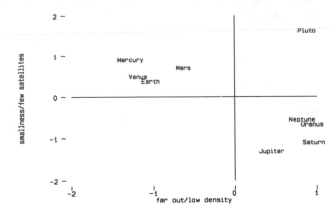

**Figure 12.2**

## Principal Factoring

Principal factoring extracts principal components from a modified correlation matrix, in which the main diagonal consists of communality estimates instead of 1's. The **factor** options **pf** and **ipf** both call for principal factoring. They differ in how communalities are estimated:

    **pf**        communality estimates equal $R^2$ from regressing each variable on all the others.

    **ipf**      iterative estimation of communalities.

Whereas principal components analysis focuses on explaining the variables' variance, principal factoring aims to explain intervariable correlations.

    Applying principal factoring with iterated communalities (**ipf**) to the planetary data:

```
. factor rings logdsun-logdense, ipf
(obs=9)
```

           (iterated principal factors; 5 factors retained)

| Factor | Eigenvalue | Difference | Proportion | Cumulative |
|--------|-----------|-----------|-----------|-----------|
| 1 | 4.59663 | 3.46817 | 0.7903 | 0.7903 |
| 2 | 1.12846 | 1.05107 | 0.1940 | 0.9843 |
| 3 | 0.07739 | 0.06438 | 0.0133 | 0.9976 |
| 4 | 0.01301 | 0.01176 | 0.0022 | 0.9998 |
| 5 | 0.00125 | 0.00137 | 0.0002 | 1.0000 |
| 6 | -0.00012 | . | -0.0000 | 1.0000 |

Factor Loadings

| Variable | 1 | 2 | 3 | 4 | 5 | Uniqueness |
|----------|------|------|------|------|------|-----------|
| rings | 0.97599 | 0.06649 | 0.11374 | -0.02065 | -0.02234 | 0.02916 |
| logdsun | 0.65708 | -0.67054 | 0.14114 | 0.04471 | 0.00816 | 0.09663 |
| lograd | 0.92670 | 0.37001 | -0.04504 | 0.04865 | 0.01662 | -0.00036 |
| logmoons | 0.96738 | -0.01074 | 0.00781 | -0.08593 | 0.01597 | 0.05636 |
| logmass | 0.83783 | 0.54576 | 0.00557 | 0.02824 | -0.00714 | -0.00069 |
| logdense | -0.84602 | 0.48941 | 0.20594 | -0.00610 | 0.00997 | 0.00217 |

    Only the first two factors have eigenvalues above 1. With **pcf** or **pf** factoring, we can simply disregard minor factors. Using **ipf**, however, we must decide how many factors to retain, then repeat the analysis asking only for that many factors:

```
. factor rings logdsun-logdense, ipf factor(2)
(obs=9)
```

```
                 (iterated principal factors; 2 factors retained)
      Factor      Eigenvalue      Difference    Proportion    Cumulative
-------------------------------------------------------------------------
        1          4.57495         3.47412        0.8061        0.8061
        2          1.10083         1.07631        0.1940        1.0000
        3          0.02452         0.02013        0.0043        1.0043
        4          0.00439         0.00795        0.0008        1.0051
        5         -0.00356         0.02182       -0.0006        1.0045
        6         -0.02537            .          -0.0045        1.0000
```

```
                   Factor Loadings
      Variable |      1           2        Uniqueness
   -----------+------------------------------------
        rings |   0.97474      0.05374      0.04699
      logdsun |   0.65329     -0.67309      0.12016
       lograd |   0.92816      0.36047      0.00858
     logmoons |   0.96855     -0.02278      0.06139
      logmass |   0.84298      0.54616     -0.00890
     logdense |  -0.82938      0.46490      0.09599
```

After this final factor analysis, we can create composite variables by `rotate` and `score`. Rotation of the `ipf` factors produces results similar to those found earlier with `pcf`: a smallness/few satellites dimension, and a far out/low density dimension. When variables have a strong factor structure, as these do, the specific techniques we choose make less difference.

## Maximum-Likelihood Factoring

The `ml` option calls for maximum-likelihood factoring:

```
. factor rings logdsun-logdense, ml factor(1)
(obs=9)
Iteration 0:   Log Likelihood =-2188.7586
Iteration 1:   Log Likelihood =-243.02794
Iteration 2:   Log Likelihood =-71.635796
Iteration 3:   Log Likelihood =-42.336298
Iteration 4:   Log Likelihood =-42.320907
Iteration 5:   Log Likelihood =-42.320558
Iteration 6:   Log Likelihood =-42.320537
Iteration 7:   Log Likelihood =-42.320535
Iteration 8:   Log Likelihood =-42.320535
Iteration 9:   Log Likelihood =-42.320535
```

```
                 (maximum-likelihood factors; 1 factor retained)
      Factor      Variance        Difference    Proportion    Cumulative
-------------------------------------------------------------------------
        1          4.47257            .           1.0000        1.0000
```

Test:  1 vs. no    factors.  Chi2(  6) =   62.02, Prob > chi2 =   0.0000
Test:  1 vs. more factors.  Chi2(  9) =   51.73, Prob > chi2 =   0.0000

```
                   Factor Loadings
      Variable |      1        Uniqueness
   -----------+---------------------
        rings |   0.98724      0.02535
      logdsun |   0.59219      0.64931
       lograd |   0.93655      0.12288
     logmoons |   0.95890      0.08052
      logmass |   0.86919      0.24451
     logdense |  -0.77145      0.40487
```

Unlike principal components or principal factoring, maximum-likelihood factor analysis supports formal tests for the appropriate number of factors. Stata's output includes two $\chi^2$ tests:

*J* vs. no factors    tests whether the current model, with *J* factors, fits the observed correlation matrix significantly better than a no-factor model. A low probability indicates that the current model is a significant improvement over no factors.

*J* vs. more factors    tests whether the current *J*-factor model fits significantly worse than a more complicated, perfect-fit model. A low *P*-value suggests that the current model does not have enough factors.

The 1-factor example above yields these results:

1 vs. no factors    probability of a greater $\chi^2 = 0.0000$ (actually, meaning $P < .00005$). The 1-factor model significantly improves upon a no-factor model.

1 vs. more factors    probability of a greater $\chi^2 = 0.0000$ ($P < .00005$). The 1-factor model is significantly worse than a perfect-fit model.

Perhaps a 2-factor model will do better:

```
. factor rings logdsun-logdense, ml factor(2)
(obs=9)
Iteration 0:   Log Likelihood =-12.540107
Iteration 1:   Log Likelihood =-11.770319
Iteration 2:   Log Likelihood = -8.456887
Iteration 3:   Log Likelihood =-6.2593382

               (maximum-likelihood factors; 2 factors retained)
     Factor      Variance      Difference     Proportion     Cumulative
------------------------------------------------------------------------
       1          3.64200        1.67115         0.6489         0.6489
       2          1.97085           .            0.3511         1.0000

Test:  2 vs. no    factors.  Chi2( 12) =  134.14, Prob > chi2 =  0.0000
Test:  2 vs. more factors.  Chi2(  4) =    6.72, Prob > chi2 =  0.1513

               Factor Loadings
 Variable |      1           2      Uniqueness
----------+------------------------------------
    rings |   0.86551    -0.41545     0.07829
  logdsun |   0.20920    -0.85593     0.22361
   lograd |   0.98438    -0.17528     0.00028
 logmoons |   0.81560    -0.49982     0.08497
  logmass |   0.99965     0.02639     0.00000
 logdense |  -0.46434     0.88565     0.00000
```

Now we find:

2 vs. more factors    probability of a greater $\chi^2 = 0.1513$. The 2-factor model is not significantly worse than a perfect-fit model.

These tests suggest that two factors provide an adequate model.

Computational routines performing maximum-likelihood factor analysis often yield "improper solutions"—unrealistic results like negative variance or zero uniqueness. When this happens (as in the 2-factor **ml** example above), the $\chi^2$ tests lack formal justification. Used cautiously, they may nonetheless provide informal guidance about the number of factors.

## Also Type help

| | |
|---|---|
| **alpha** | Cronbach's α reliability |
| **factor** | extract initial factors or principal components |
| **greigen** | scree graph of eigenvalue |
| **rotate** | rotate factor matrix (after **factor**) |
| **score** | generate factor scores (after **factor** or **rotate**) |

# 13
# List of Some Stata Commands

This chapter lists many of Stata's data-handling and analytical commands. It is not a complete list—the *Stata Reference Manual* takes three volumes. Furthermore, Stata's capabilities grow bimonthly via the *Stata Technical Bulletin*. Many new commands will exist by the time you read this. The list also omits many features useful in writing Stata programs. For guidance on Stata programming, study ado-files as examples, and consult the *Stata Reference Manual*.

The list is grouped by broad topic areas, and alphabetically within area. Some procedures appear more than once. Unless noted otherwise, we can summon online help regarding any command by typing **help** followed by the command's name.

## Data and Variables

| | |
|---|---|
| append | append two datasets (increasing *n* of observations) |
| cf | compare two files |
| collapse | make dataset of means, medians, etc. |
| compare | compare two variables |
| compress | compress data in memory |
| convert | convert raw data into cross-product data |
| count | count observations satisfying specified condition |
| cross | form every pairwise combination of two datasets |
| dates | date conversions |
| describe | describe contents of data in memory or on disk |
| drop | eliminate variables or observations |
| egen | extended **generate** |
| encode | encode string variables into numeric and vice versa |
| expand | duplicate observations |
| fillin | rectangularize dataset |
| format | specify permanent display format |
| generate | create variable from expression |
| infile | read text (ASCII) file into memory |
| input | enter data from keyboard |
| inspect | simple summary of variable's characteristics |
| label | label manipulation (data, variable, values) |
| list | list values of variable |
| merge | merge two datasets (increasing number of variables) |
| modify | interactively modify data values |
| mvencode | change missing to coded missing value and vice versa |
| order | reorder variables in dataset |
| outfile | write ASCII-format dataset |
| recast | change storage type of variable |
| recode | recode categorical variable |
| rename | rename variable |
| replace | change contents of variable |

| | |
|---|---|
| reshape | convert data from wide to long and vice versa |
| sample | draw random sample |
| save | save a dataset to disk |
| sort | sort data |
| stack | stack data vertically |
| use | retrieve a dataset from disk |
| xpose | interchange observations and variables |

## Diagnostic Statistics

| | |
|---|---|
| fpredict | diagnostics after fit; see pages 106–107 (help fit) |
| lpredict | diagnostics after logistic; see page 144 (help logistic) |
| predict | residuals and diagnostics after regress, anova, etc.; see page 75 |

## egen (Extended generate) Functions

| | |
|---|---|
| count | count the number of nonmissing values of expression (help egen) |
| diff | equals 1 when variables in varlist unequal, 0 otherwise (help egen) |
| iqr | interquartile range of expression (help egen) |
| ma | span-# moving average of expression (help egen) |
| max | maximum value of expression (help egen) |
| mean | mean of expression (help egen) |
| median | median of expression (help egen) |
| min | minimum of expression (help egen) |
| pctile | #th percentile of expression (help egen) |
| rank | creates ranks of expression (help egen) |
| rmean | row mean of varlist (help egen) |
| rmiss | number of missing variables in varlist (help egen) |
| robs | number of nonmissing variables in varlist (help egen) |
| rsum | row sum of varlist (help egen) |
| sd | standard deviation of expression (help egen) |
| std | standardized values of expression (help egen) |
| sum | sum of expression (help egen) |

## Epidemiological Tables

| | |
|---|---|
| ir | incidence rate data—estimates, confidence intervals from data (help epitab) |
| iri | incidence rate data—estimates, intervals from statistics (help epitab) |
| cs | cohort study data—estimates, intervals from data (help epitab) |
| csi | cohort study data—estimates, intervals from statistics (help epitab) |
| cc | case-control and cross-sectional data—odds ratio etc. from data (help epitab) |
| cci | case-control and cross-sectional data—odds ratio etc. from statistics (help epitab) |
| mcc | matched-case control data—McNemar's $\chi^2$, etc. from data (help epitab) |
| mcci | matched-case control data—McNemar's $\chi^2$, etc. from statistics (help epitab) |

# Functions

For listings of mathematical and statistical functions for `generate`, `replace`, or `display` commands, see pages 26–28; also `help functions`

# Graphs

| | |
|---|---|
| `acprplot` | Mallows' augmented component-plus-residual (partial residual) plot (`help fit`) |
| `avplot` | added-variable plot or partial regression leverage plot (`help fit`) |
| `avplots` | all added-variable plots (`help fit`) |
| `cchart` | $c$ chart (`help qc`) |
| `coxbase` | baseline survival curve (`help cox`) |
| `cprplot` | component-plus-residual or partial residual plot (`help fit`) |
| `dydx` | derivative of numerical function (`help range`) |
| `egen` | generate running averages for smoothing |
| `gphdot` | dot-matrix or laser printing |
| `gphpen` | pen plotting, PostScript, or change to pic format |
| `graph` | general graphing command |
| `greigen` | eigenvalue or scree graph |
| `gr3` | 3-dimensional scatterplot |
| `gwood` | Kaplan-Meier curve w/Greenwood limits (`help survival`) |
| `integ` | integrate numerical function (`help range`) |
| `kapmeier` | Kaplan-Meier survival curve (`help survival`) |
| `ksm` | lowess smoothing |
| `loglogs` | log-log survival Weibull check (`help survival`) |
| `lroc` | graph and calculate area under ROC curves for model estimated by `logistic` (`help logistic`) |
| `lvr2plot` | leverage versus squared residual or L-R plot (`help fit`) |
| `pchart` | $p$ chart—quality control (`help qc`) |
| `plot` | text-mode scatterplots |
| `pnorm` | standardized normal probability plot (`help diagplots`) |
| `qnorm` | quantile-normal plots (`help diagplots`) |
| `qqplot` | quantile-quantile plots (`help diagplots`) |
| `quantile` | quantile plots (`help diagplots`) |
| `range` | generate numerical range for graphing |
| `rchart` | $R$ chart—quality control (`help qc`) |
| `rvfplot` | residual versus fitted, or e versus $Y$-hat plot (`help fit`) |
| `rvpplot` | residual versus predictor plot (`help fit`) |
| `serrbar` | standard-error bar chart |
| `shewhart` | $X$-bar and $R$ chart in same image (`help qc`) |
| `stem` | stem-and-leaf displays |
| `symplot` | symmetry plots (`help diagplots`) |
| `window` | graphics windows—Unix versions only |
| `xchart` | $X$-bar chart—quality control (`help qc`) |

# Miscellaneous Statistics

| | |
|---|---|
| `alpha` | Cronbach's $\alpha$ reliability |

| | |
|---|---|
| boot | bootstrap sampling |
| boxcox | Box-Cox transformation toward normality |
| ci | confidence intervals for means, proportions, and counts |
| correlate | correlation and covariance matrices, for variables and coefficients |
| cumul | cumulative distribution |
| deff | Kish design effect |
| fpredict | regression diagnostics (help fit) |
| huber | Huber jackknife variance estimates |
| impute | predict missing values |
| kappa | interrater agreement |
| ktau | Kendall's $\tau$ rank correlation (help spearman) |
| ladder | ladder of powers |
| lnskew0 | find zero-skewness log or Box-Cox transformation |
| lpredict | logistic regression diagnostics (help logistic) |
| lv | letter-value displays |
| means | arithmetic, geometric, and harmonic means |
| pcorr | partial correlation coefficients |
| predict | obtain residuals, predictions, diagnostics after model estimation |
| range | numerical ranges, derivatives, and integrals |
| spearman | Spearman rank correlation |
| summarize | univariate summary statistics |
| tabulate | one- and two-way tables of frequencies or means |

## Model Fitting

| | |
|---|---|
| anova | $K$-way ANOVA, ANCOVA, and regression |
| blogit | ML estimation of logit regression with grouped (blocked) data (help glogit) |
| bprobit | ML estimation of probit regression with grouped (blocked) data (help glogit) |
| cnreg | censored normal regression |
| constrain | define, list, or drop linear constraints (help constrain) |
| corc | Cochrane-Orcutt regression, corrected for serially correlated residuals |
| cox | Cox (nonparametric survival) regression |
| ereg | exponential (parametric survival) regression (help weibull) |
| estimate | hold results from recent model-fitting in memory |
| factor | principal components and factor analysis |
| fit | OLS with diagnostic statistics and graphs |
| glogit | minimum $\chi^2$ estimation of logit regression with grouped data |
| gprobit | minimum $\chi^2$ estimation of probit regression with grouped data (help glogit) |
| hlogit | logit regression with Huber jackknife standard errors |
| hprobit | probit regression with Huber jackknife standard errors (help hreg) |
| hreg | OLS regression with Huber jackknife standard errors |
| impute | predict missing values by best-subsets regression |
| linktest | specification link test for single-equation models |
| logistic | logistic regression (dichotomous $Y$) giving odds ratios and diagnostics |
| logit | logistic regression (dichotomous $Y$) giving coefficients |
| loneway | large oneway ANOVA, random effects, and reliability |

| | |
|---|---|
| mlogit | multinomial logit |
| ologit | logistic regression with ordered-category $Y$ |
| oneway | oneway ANOVA, including multiple-comparison tests |
| oprobit | probit regression with ordered-category $Y$ |
| poisson | Poisson regression |
| probit | probit regression with dichotomous $Y$ variable |
| qreg | quantile (including median) regression |
| regdw | OLS regression with Durbin-Watson statistic (help corc) |
| regress | OLS regression; also WLS and instrumental variables |
| rotate | rotate factor matrix |
| rreg | robust regression, Huber and biweight IRLS |
| score | generate factor scores |
| tobit | tobit regression |
| weibull | Weibull (parametric survival) regression |

## Quality Control

| | |
|---|---|
| cchart | $c$ chart (help qc) |
| pchart | $p$ (fraction defective) chart (help qc) |
| rchart | $R$ (range or dispersion) chart (help qc) |
| shewchart | vertically aligned $X$-bar and $R$ charts (help qc) |
| xchart | $X$-bar (control line) chart (help qc) |

## Stepwise Variants of Model Fitting Procedures

| | |
|---|---|
| stepwise | stepwise regress |
| swcnreg | stepwise cnreg |
| swcox | stepwise cox |
| swereg | stepwise ereg |
| swlogis | stepwise logistic |
| swlogit | stepwise logit |
| swologit | stepwise ologit |
| swoprbt | stepwise oprobit |
| swpois | stepwise poisson |
| swprobit | stepwise probit |
| swqreg | stepwise qreg |
| swtobit | stepwise tobit |
| swweib | stepwise weibull |

## Survival Analysis (Event History Analysis)

| | |
|---|---|
| cox | Cox regression, proportional hazards model |
| ereg | exponential regression, parametric survival model (help weibull) |
| gwood | graph Kaplan-Meier survival curve with Greenwood confidence bands (help survival) |
| kapmeier | graph Kaplan-Meier survival curve (help survival) |
| loglogs | graph log($-$log($s$)) vs. log(time), where $s$ is K-M survival estimate (help survival) |

| logrank | log-rank test of equality of two or more survival curves (**help survival**) |
| ltable | life tables for survival data |
| mantel | Mantel-Haenszel test comparing two survival curves (**help survival**) |
| survcurv | K-M survival estimate, Greenwood s.d., variance log survival (**help survival**) |
| survsum | survival time summary statistics (**help survival**) |
| weibull | Weibull regression |
| wilcoxon | Gehen generalization of Wilcoxon test for equality of two survival distributions (**help survival**) |

## Tests

| bitest | exact binomial probabilities test from data |
| bitesti | exact binomial probabilities test from summary statistics (**help bitest**) |
| ksmirnov | Kolmogorov-Smirnov equality of distributions test |
| kwallis | Kruskal-Wallis equality of medians test |
| linktest | specification link test for single-equation models |
| lrtest | likelihood-ratio tests after ML model fitting |
| ranksum | Wilcoxon rank-sum (Mann-Whitney $U$) two-sample test (**help signrank**) |
| sdtest | test equality of standard deviations (variances) from data |
| sdtesti | test equality of standard deviations (variances) from statistics (**help sdtest**) |
| signrank | Wilcoxon matched-pairs signed ranks test |
| signtest | test equality of medians for matched pairs of observations (**help signrank**) |
| sktest | skewness and kurtosis test of normality |
| tabulate | $\chi^2$ tests and correlations for crosstabulation |
| test | user-specified hypothesis tests after model fitting |
| ttest | one- and two-sample $t$ tests from data |
| ttesti | one- and two-sample $t$ tests from summary statistics (**help ttest**) |

## Other *Stata Technical Bulletin* Commands Included with Student Stata

| breg | bounded-influence robust regression (STB-2) |
| centile | centile estimation (STB-8) |
| discrim | dichotomous discriminant analysis (STB-5) |
| friedman | Friedman's ANOVA and Kendall's coefficient of concordance (STB-3) |
| loglin | loglinear analysis (STB-8) |
| lwald | standardized coefficients, wald statistics, and partial correlations following logistic regression (STB-8) |
| manova | oneway multivariate ANOVA (STB-6) |
| multnorm | test for multivariate normality (STB-2) |
| nl | nonlinear regression (STB-7) |
| nlsm | nonlinear, robust smoother (STB-8) |
| ranova | repeated measures ANOVA (STB-4) |
| unblock | unblocks grouped datasets (STB-7) |

# 14
# Monte Carlo and Bootstrap Methods

Monte Carlo methods involve hundreds, thousands, or even millions of times more calculation than the traditional statistics described elsewhere in this book. By analyzing many computer-generated random samples, we learn about the long-run performance of our statistical techniques. Such computer experiments (called Monte Carlo for their chance or "gambling" aspect) have dramatically extended the frontiers of statistical knowledge. Monte Carlo research can explore problems too complex to solve by mathematical reasoning. For other problems, with already well-known solutions, computer-intensive methods provide a new teaching tool.

Because it demands so much calculation, Monte Carlo work until recently meant mainframe computers—and even there, it was costly. Stata proves to be surprisingly capable, however. The slower pace of desktop computing is offset by the ease and speed of using Stata to investigate the sometimes surprising analytical results.

This chapter begins with some basics of random data generation. It then shows two simple Monte Carlo experiments, that you can run as is or modify. Finally, we look at the related topic of bootstrapping. Because computer-intensive research often requires writing new programs, this chapter also provides a glimpse of Stata programming.

## Random Number Generation

Stata's random-number function, `uniform()`, adapts to generate random data from a variety of theoretical distributions. We can add random variables to an existing dataset through `generate` or `replace`, or we can create artificial datasets from scratch. For example:

```
. clear
. set obs 150
. generate newvar = uniform()
```

`set obs 150` tells Stata the new dataset will contain $n = 150$ observations. `generate newvar = uniform()` then creates a variable called *newvar*, containing 150 (16-digit) random numbers from a uniform distribution over the open interval from 0 to 1.

Chapter 2 described `uniform()`, Stata's random-number function. If we have not yet used `uniform()` during our current Stata session, and we issue the command `generate newvar = uniform()`, the first five *newvar* values are:

```
. list in 1/5

        newvar
  1.    .0321949
  2.    .6710823
  3.    .9830985
  4.    .6296493
  5.    .2459989
```

Stata ordinarily uses 1001 as the seed to generate a session's first random number. After an initial seed, each pseudo-random number derives from the one before, as described on page 29. Starting again from the same seed therefore guarantees an identical stream of numbers. We can change the seed using a `set seed` command, specifying any large odd number such as:

```
. set seed 1111
```

In research, we may want to set either the same seed as last time (to generate the same sequence) or a different seed (to ensure a different sequence), depending on our needs.

If we want *newvar* to follow a uniform distribution over {0, 428} instead of {0, 1}:

```
. generate newvar = 428*uniform()
```

These will still be 16-digit numbers. Perhaps we want only integers from 1 to 428:

```
. generate newvar = 1 + int(428*uniform())
```

Random numbers can simulate a variety of chance processes. For example, to simulate 100 throws of a six-sided die:

```
. clear
. set obs 100
. generate die = 1 + int(6*uniform())
. tabulate die
```

| die | Freq. | Percent | Cum. |
|---|---|---|---|
| 1 | 21 | 21.00 | 21.00 |
| 2 | 13 | 13.00 | 34.00 |
| 3 | 13 | 13.00 | 47.00 |
| 4 | 14 | 14.00 | 61.00 |
| 5 | 22 | 22.00 | 83.00 |
| 6 | 17 | 17.00 | 100.00 |
| Total | 100 | 100.00 | |

We might expect to see 16.67% ones, 16.67% twos, and so on, but in any one sample, like these 100 "throws," observed percentages will vary randomly around theoretical values. (With this and other random-data exercises, your own results will probably differ from the examples I give.)

Simulating 100 throws of a pair of six-sided dice:

```
. generate dice = 2 + int(6*uniform()) + int(6*uniform())
. tabulate dice
```

| dice | Freq. | Percent | Cum. |
|---|---|---|---|
| 2 | 1 | 1.00 | 1.00 |
| 3 | 5 | 5.00 | 6.00 |
| 4 | 8 | 8.00 | 14.00 |
| 5 | 9 | 9.00 | 23.00 |
| 6 | 14 | 14.00 | 37.00 |
| 7 | 13 | 13.00 | 50.00 |
| 8 | 13 | 13.00 | 63.00 |
| 9 | 21 | 21.00 | 84.00 |
| 10 | 9 | 9.00 | 93.00 |
| 11 | 5 | 5.00 | 98.00 |
| 12 | 2 | 2.00 | 100.00 |
| Total | 100 | 100.00 | |

## "Resampling Statistics"

Under the name "resampling statistics," computer simulation methods have recently been advocated as a nonmathematical way to approach almost any statistical problem, and particularly as an aid in teaching (Simon and Bruce 1991a, 1991b). Two elementary examples:

1.  Suppose a couple has three children. What is the probability that at least one will be a girl? For simplicity, assume that boy and girl births are equally likely.

a.   Theoretical solution: there are eight equally probable outcomes (BBB, BBG, BGB, etc.), seven of which include girls, so the answer is 7/8 = .875. More formal thinkers might apply the Binomial Theorem, or ask Stata to do so. To find the probability of 1 or more successes, in 3 trials, where the probability of success in a single trial is .5:

```
. display Binomial(3,1,.5)
.875
```

b.   Resampling approximation: we create a variable simulating the number of girls in 100 families of three, and examine its frequency distribution:

```
. clear
. set obs 100
. generate ngirls = int(2*uniform())+int(2*uniform())+int(2*uniform())
. tabulate ngirls, plot
```

```
ngirls|        Freq.
--------+------------+----------------------------------------------------
     0 |         15  |*****************
     1 |         40  |******************************************************
     2 |         29  |*************************************
     3 |         16  |********************
--------+------------+----------------------------------------------------
 Total |        100
```

In this experiment, 85% of the "families" had at least one girl (suggesting $P = .85$; compare with theoretical value .875).

2.   A certain basketball player usually scores on 47% of his shots. If this probability does not change during the game (an issue of some controversy among statisticians and sports fans), how likely is he to miss at least eight of his first ten shots?

a.   Theoretical solution: again the Binomial Theorem applies. The probability of missing is .53, so the probability of 8 or more misses in 10 trials equals about .079:

```
. display Binomial(10, 8, .53)
.07914701
```

b.   Resampling approximation: we create 10 dummy variables (*hit1*, *hit2*, . . . *hit10*) to indicate whether each shot hit or missed (1 or 0), then add them to get the number of hits:

```
. clear
. set obs 100
. generate hit1 = uniform() <.47
. generate hit2 = uniform() <.47
. generate hit3 = uniform() <.47
.

.
. generate hit10 = uniform() <.47
. egen hits = rsum(hit1-hit10)
. tabulate hits, plot
```

```
   hits|      Freq.
--------+-----------+-------------------------------------------------------
      1 |         1 |**
      2 |         9 |*****************
      3 |        11 |*********************
      4 |        26 |******************************************************
      5 |        22 |*************************************************
      6 |        12 |***********************
      7 |        12 |***********************
      8 |         6 |************
      9 |         1 |**
--------+-----------+-------------------------------------------------------
  Total |       100
```

This experiment found 2 or fewer hits in 10 of our 100 imaginary "games" (suggesting $P = .10$; compare with theoretical value .079).

In these examples, the theoretical solution is correct, whereas resampling just provides an approximation. If we kept on repeating a resampling experiment, each time with new random data, we would get a variety of different outcomes—sometimes far from the true (theoretical, or infinite-sample) answer. For many people, though, the resampling approach seems easier to understand. Its precision should improve if we draw more samples—perhaps 1,000 or 10,000 repetitions instead of 100.

## Sampling from Theoretical Probability Distributions

A standard normal distribution, N(0, 1), has mean $\mu = 0$ and standard deviation $\sigma = 1$. We can **generate** a variable containing random values from a standard normal distribution:

```
. generate Z1 = invnorm(uniform())
```

Values from a normal distribution with $\mu = 500$ and $\sigma = 75$:

```
. generate X1 = 500+75*invnorm(uniform())
```

Two standard normal variables with population correlation $\rho = .8$:

```
. generate Z1 = invnorm(uniform())
. generate Z2 = Z1*.8 + invnorm(uniform())*sqrt(1-.8^2)
```

The .8 in this example could be replaced with any other correlation desired.

After they are generated, variables can be moved to another mean and standard deviation without affecting their correlation. For example, to simulate $Z1 \sim N(500, 75)$ and $Z2 \sim N(80, 10)$, with correlation $\rho = .8$, first **generate** as above. Then change the means and standard deviations:

```
. replace Z1 = 500 + 75*Z1
. replace Z2 = 80 + 10*Z2
```

Their theoretical correlation still equals .8. Random-variable generation imitates random sampling from a population with the specified parameters. Sample means, correlations, etc. will not exactly equal the theoretical parameters.

If $Z$ follows a normal distribution, $L = e^Z$ follows a lognormal distribution. To form a lognormal variable $L$ based upon a standard normal distribution:

```
. generate L = exp(invnorm(uniform()))
```

To form lognormal variable $L$, based on an N(5, 2) distribution:

```
. generate L = exp(5 + 2*invnorm(uniform()))
```

Taking logarithms, of course, normalizes lognormal variables.

$E$ is drawn randomly from an exponential distribution with mean and standard deviation $\mu = \sigma = 3$:

```
. generate E = -3*ln(uniform())
```

For other means and standard deviations, substitute other values for 3.

A $\chi^2$ (chi-square) variable with $df$ degrees of freedom (mean $\mu = df$ ; variance $\sigma^2 = 2df$ ) equals the sum of $df$ independent, squared standard normal variables. For example, to generate values from a $\chi^2$ distribution with one degree of freedom:

```
. generate X1 = (invnorm(uniform()))^2
```

From a $\chi^2$ distribution with two degrees of freedom:

```
. generate X2 = (invnorm(uniform()))^2 + (invnorm(uniform()))^2
```

and so forth. Like exponential and lognormal distributions, $\chi^2$ distributions are positively skewed and include no negative values. The skewness of $\chi^2$ lessens with increasing degrees of freedom.

If $Z$ is a standard normal variable and $X2$ is a $\chi^2$ variable, independent of $Z$ and with $df$ degrees of freedom, then the ratio $Z / \sqrt{X2/df}$ follows a Student's $t$ distribution with $df$ degrees of freedom. We could use this definition to produce random $t$ values, but an easier approach employs Stata's inverse $t$ function `invt()`. To generate random values from a Student's $t$ distribution with $df = 10$:

```
. generate t = invt(10, uniform())
```

Like the standard normal distribution, $t$ distributions are symmetrical, centered on zero, and range from positive to negative infinity. $t$ distributions have heavier-than-normal tails, most noticeable for low $df$.

The ratio of two independent $\chi^2$ variables, each divided by its degrees of freedom, follows an $F$ distribution. If $X1$ is distributed as $\chi^2 [df_1]$, and $X2$ is distributed as $\chi^2 [df_2]$, then the ratio $(X1/df_1) / (X2/df_2)$ follows an $F$ distribution with $df_1$ and $df_2$ degrees of freedom. For example, to generate values from $F(3,12)$, first obtain $\chi^2$ variables $X3$ (sum of three squared $N(0, 1)$ variables) and $X12$ (sum of 12 squared $N(0, 1)$ variables) as described above. Then:

```
. generate F = (X3/3)/(X12/12)
```

$F$ distributions contain only positive values.

Contaminated distributions mix two or more simpler distributions. For example:

$$Y \sim N(0, 1) \quad \text{with probability .95}$$
$$N(0, 10) \quad \text{with probability .05}$$

Thus for 95% of the population, $Y$ follows a standard normal distribution. For the remaining 5%, though, $Y$ follows a normal distribution with much wider spread—a standard deviation of 10 instead of 1. Contaminated distributions, simulating data with a small proportion of wild errors, are often used in robustness research to test resistance to gross outliers. To draw samples from such a distribution:

```
. generate Y = invnorm(uniform())
. replace Y = 10*Y if uniform()<.05
```

We first generate $Y$ as a standard normal variable. The second line multiplies $Y$ by 10 (corresponding to a tenfold increase in standard deviation) for a randomly-selected subset of cases. Over the long run, about 5% of the cases get modified.

## Monte Carlo Example 1:  The Central Limit Theorem

To perform a Monte Carlo experiment with Stata, we need a program (do-file) that:
1.   generates a random sample according to our chosen model, analyzes this sample, and writes the results to a log file;
2.   repeats step 1 many times, then closes the log file; and
3.   uses **infile** to read the log file, creating a new dataset of the experimental results.
This section gives a simple example.

The Central Limit Theorem asserts that, as sample size becomes large, the sampling distribution of the mean tends toward normality (regardless of the variable's distribution), centered on the population mean μ, and with standard error $\sigma / \sqrt{n}$ . This result derives from mathematical reasoning, but we can explore how it "really works" using Monte Carlo techniques.

Program *monte1.do* generates samples consisting of $n = 100$ random $Y$ values, drawn from an exponential distribution with $\mu = \sigma = 2$. It uses **summarize** to find the mean of each sample, writes that mean (and the sample number) in a log file, then **clear**s memory and creates another random sample. To stay within the limits of Student Stata, *monte1.do* draws only 160 random samples. For serious research, 1,000 or more samples would be better.

```
program define monte1
    version 3.0
    set seed 1111
    set more 1
    capture erase c:\stustata\monte1.log
    log using c:\stustata\monte1.log
    local sample=1
    tempvar Y
        while 'sample'<161 {
                quietly   {
                        clear
                        set obs 100
                        generate 'Y'=-2*ln(uniform())
                }
                quietly summ 'Y'
                display 'sample'
                display _result(3)
                local sample='sample'+1
        }
        log close
        clear
        infile sample meanY using c:\stustata\monte1.log
        label data "MC experiment, n=100, mu=2"
        label variable sample "sample number"
        label variable meanY "mean of exponential(2) Y"
        save c:\stustata\monte1, replace
end
```

Programs like *monte1.do* can be written with any word processor, and saved as an ASCII (text) file. If *monte1.do* resides in directory *c:\stustata*, then to run it tell Stata:

```
. run c:\stustata\monte1.do
. monte1
```

If you make your own modifications to *monte1.do* and then want to run it again, first **drop** the old program from memory:

```
. program drop _all
. run c:\stustata\monte1.do
. monte1
```

*montel.do* records means from each artificial sample in log file *montel.log*. This log file, and the final data file, also go in directory *c:\stustata*. If you are not using this location, change the appropriate lines in *montel.do* before you **run** it. After 160 samples, an **infile** statement reads *montel.log*, labels the results, and saves file *montel.dta* containing 160 means:

```
. describe

Contains data from c:\stustata\monte1.dta
  Obs:     160 (max=  2620)              MC experiment, n=100, mu=2
  Vars:      2 (max=    99)
Width:       8 (max=   200)
  1. sample         float   %9.0g        sample number
  2. meanY          float   %9.0g        mean of exponential(2) Y
Sorted by:

. summarize

Variable |     Obs        Mean    Std. Dev.       Min        Max
---------+-----------------------------------------------------------
  sample |     160        80.5    46.33213          1        160
   meanY |     160    1.995626    .1801956   1.558867   2.532398
```

The mean of our 160 sample means equals 1.995626, close to the true population mean μ = 2. The standard deviation of our 160 sample means equals .1801956, not far from the Central Limit Theorem's prediction $\sigma / \sqrt{n} = 2 / \sqrt{100} = 0.2$.

The Central Limit Theorem also predicts that the distribution of sample means tends towards normality. *montel.do* drew the original *Y* values from an exponential(2) distribution, illustrated in Figure 14.1 with a sample of 10,000 observations. Professional Stata produced this graph:

```
. clear
. set maxobs 10000
. set obs 10000
. generate Y = -2*ln(uniform())
. graph Y, xlabel ylabel bin(12) b2(Exponential Variable with Mean = 2)
```

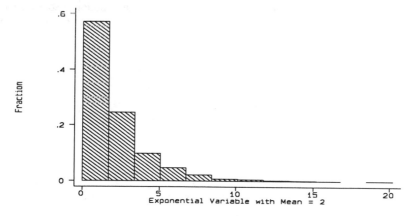

**Figure 14.1**

Although the distribution of *Y* is radically nonnormal (Figure 14.1), the Monte Carlo sampling distribution of *meanY* conforms reasonably well to a theoretical normal curve (Figure 14.2):

```
. use c:\stustata\monte1, clear
. graph meanY, bin(12) norm(2,.2) xlabel b2(sample mean)
```

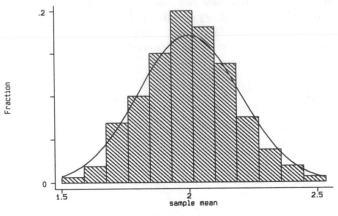

**Figure 14.2**

The contrast between Figures 14.1 and 14.2 illustrates not only the Central Limit Theorem, but also the conceptual distinction between the distribution of a variable across cases (Figure 14.1) and the sampling distribution of a statistic across samples (Figure 14.2). The next section provides another Monte Carlo example, demonstrating a way to evaluate the efficiency of competing statistical techniques.

## Monte Carlo Example 2:  Comparing Median and Mean

Program *monte2.do* experimentally compares the performance of the mean and median, applied to two variables, $X$ and $W$, in 150 random samples of $n = 100$ values each.  $X$ follows a standard normal distribution with mean 0, standard deviation 1.  $W$ follows a contaminated normal distribution:  90% N(0, 1), and 10% N(0, 10).  Though $W$'s distribution also has 0 mean, it includes a substantial fraction of wild outliers that appear unpredictably in random samples.

```
program define monte2
    version 3.0
    set seed 2111
    set more 1
    capture erase c:\stustata\monte2.log
    log using c:\stustata\monte2.log
    local sample=1
    tempvar X W
        while 'sample'<151 {
                quietly  {
                        clear
                        set obs 100
                        generate 'X' = invnorm(uniform())
                        generate 'W' = 'X'
                        replace 'W'= 10*'W' if uniform()<.10
                }
                quietly summ 'X', detail
                display 'sample'
                display _result(3)
                display _result(10)
                quietly summ 'W', detail
                display _result(3)
                display _result(10)
                display
                local sample='sample'+1
        }
```

```
        log close
        clear
        infile sample meanX medianX meanW medianW using c:\stustata\monte2.log
        label data "MC experiment, n=100, mu=0"
        save c:\stustata\monte2, replace
end
```

If *monte2.do* resides in directory *c:\stustata*, then to run it tell Stata:

```
. run c:\stustata\monte2.do
. monte2
```

*monte2.do* creates a log file, *monte2.log*, that records sample number, mean *X*, median *X*, mean *W*, and median *W* for each artificial sample. As written, the program stores both *monte2.log* and *monte2.dta* in directory *c:\stustata*.

After 150 iterations, an **infile** statement reads *monte2.log*, labels the results, and saves file *monte2.dta* containing 150 means and medians:

```
. describe
```

```
Contains data from c:\stustata\monte2.dta
  Obs:    150 (max=  2620)
  Vars:     5 (max=    99)              MC experiment, n=100, mu=0
Width:     20 (max=   200)
  1. sample      float   %9.0g          sample number
  2. meanX       float   %9.0g          mean of N(0,1) X
  3. medianX     float   %9.0g          median of N(0,1) X
  4. meanW       float   %9.0g          mean of 10% contam.N(0,10) W
  5. medianW     float   %9.0g          median of 10% contam.N(0,10) W
Sorted by:
```

With enough random samples, such experiments will show that:

1.  Mean and median are both unbiased estimators of the population means of *X* and *W*. That is, the means of *meanX* and *meanW*, and the means of *medianX* and *medianW*, should all be close to $\mu = 0$.

2.  Applied to a normally distributed variable like *X*, the mean estimates $\mu$ more efficiently than does the median. That is, the standard deviation of *meanX* tends to be lower than the standard deviation of *medianX*. (Theoretically, the sampling variance of the median equals $\pi/2 \approx 1.57$ times the variance of the mean, when both are applied to large samples from a normal population.)

3.  On the other hand, applied to a heavy-tailed distribution like *W*'s contaminated normal, the mean becomes much less efficient than the median. That is, the standard deviation of *meanW* tends to be substantially higher than the standard deviation of *medianW*.

Our *monte2.do* results generally agree with these predictions:

```
. summarize
```

| Variable | Obs | Mean | Std. Dev. | Min | Max |
|---|---|---|---|---|---|
| sample | 150 | 75.5 | 43.44537 | 1 | 150 |
| meanX | 150 | .0058459 | .1135642 | -.2921029 | .2760522 |
| medianX | 150 | .0026692 | .1332964 | -.3143041 | .3775555 |
| meanW | 150 | .0126307 | .3441843 | -1.057949 | .8831863 |
| medianW | 150 | .0028854 | .1460447 | -.342418 | .3871394 |

Given normal data ($X$), the mean exhibits less sample-to-sample variation than the median. Given data from a heavy-tailed distribution ($W$), the mean's standard deviation triples, while the median's standard deviation increases less than 10%.

Figure 14.3 shows these Monte Carlo sampling distributions as boxplots. The wide variation of mean $W$ values demonstrates how the mean breaks down (but the median does not) when faced with heavy-tailed distributions.

```
. graph meanX-medianW, b2("Mean X      Median X     Mean W       Median W")
       ylabel yline(0) box l1(Sample mean)
```

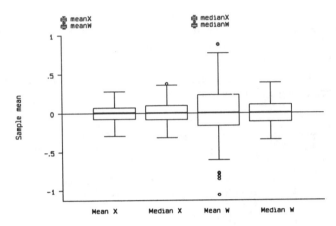

**Figure 14.3**

## Regression Models

To generate artificial data following a known regression model:

1.  Clear memory and specify sample size, then generate the $X$ variables. Suppose we want a sample of $n = 150$ and a model involving two $X$ variables, bivariate standard normal and correlated $\rho = .5$:

```
. clear
. set obs 150
. generate X1 = invnorm(uniform())
. generate X2 = X1*.5 + invnorm(uniform())*sqrt(1-.5^2)
```

2.  Calculate predicted $Y$ values ($Yhat$) according to the desired regression model. For example, choosing the model

$$E[Y] = \beta_0 + \beta_1 X1 + \beta_2 X2,$$

with $\beta_0 = 9$, $\beta_1 = 2$, and $\beta_2 = 3$:

```
. generate Yhat = 9 + 2*X1 + 3*X2
```

The theoretical variance of $Yhat$ then is:

$$\text{Var}[Yhat] = \beta_1^2 \text{Var}[X1] + \beta_2^2 \text{Var}[X2] + 2\beta_1 \beta_2 \text{Cov}[X1, X2] \qquad [14.1]$$

If $X1$ and $X2$ happen to be standardized variables (unit variance, and hence covariance equal to their correlation), the variance of $Yhat$ simplifies to

$$\mathrm{Var}[Yhat] = \beta_1^2 + \beta_2^2 + 2\beta_1 \beta_2 \rho \qquad\qquad\qquad [14.2]$$

In this example:

$$\mathrm{Var}[Yhat] = 2^2 + 3^2 + (2 \times 2 \times 3 \times 0.5)$$
$$= 19$$

3.  Generate sample $Y$ values by adding an error term ($e$) to $Yhat$. For standard normal, independent and identically distributed (normal i.i.d.) errors:

```
. generate e = invnorm(uniform())
```

This produces errors with a theoretical standard deviation ($\sigma_e$) equal to one. The model $R^2$ can be controlled by changing $\sigma_e$. Substitute the desired $R^2$, and $\mathrm{Var}[Yhat]$ from [14.1] or [14.2], into

$$\sigma_e = \sqrt{\mathrm{Var}[Yhat]/R^2 - \mathrm{Var}[Yhat]} \qquad\qquad [14.3]$$

to find the necessary value of $\sigma_e$. For example, if we want $R^2 = .7$, then given $\mathrm{Var}[Yhat]$ = 19, the errors' standard deviation must be:

$$\sigma_e = \sqrt{19/.7 - 19}$$
$$= 2.854$$

Two further commands achieve this, if we already have a variable $e \sim N(0, 1)$:

```
. replace e = e*2.854
. generate Y = Yhat + e
```

These steps should produce a sample of 150 cases, with $X1$ and $X2$ from $N(0, 1)$ populations, and $e$ from an $N(0, 2.854)$ population. **summarize** shows that sample means and standard deviations are not far from the population parameters:

```
. summarize
```

| Variable | Obs | Mean | Std. Dev. | Min | Max |
|---|---|---|---|---|---|
| X1 | 150 | .0545526 | 1.00336 | -2.740156 | 2.352067 |
| X2 | 150 | .074972 | .994774 | -2.363404 | 2.739927 |
| Yhat | 150 | 9.334021 | 4.336381 | -1.279775 | 21.49181 |
| e | 150 | .0249806 | 2.890174 | -6.593187 | 8.329257 |
| Y | 150 | 9.359002 | 5.104886 | -4.080968 | 24.93684 |

$X1$ and $X2$ have a theoretical correlation of $\rho = .5$. $e$ should be uncorrelated with either $X$ variable. Again, our sample is not far from these values:

```
. correlate
(obs=150)
```

| | X1 | X2 | Yhat | e | Y |
|---|---|---|---|---|---|
| X1 | 1.0000 | | | | |
| X2 | 0.4902 | 1.0000 | | | |
| Yhat | 0.8001 | 0.9150 | 1.0000 | | |
| e | -0.0526 | -0.0283 | -0.0438 | 1.0000 | |
| Y | 0.6499 | 0.7613 | 0.8247 | 0.5290 | 1.0000 |

Finally, $Y$ theoretically obeys the linear relation:

$$E[Y] = 9 + 2X1 + 3X2$$

with $X1$ and $X2$ explaining 70% of $Y$'s variance. **regress** confirms that this sample resembles the theoretical model:

```
. regress Y X1 X2

    Source |       SS       df       MS                    Number of obs =      150
-----------+------------------------------                 F(  2,   147) =   156.44
     Model | 2641.75924      2 1320.87962                  Prob > F      =   0.0000
  Residual | 1241.16014    147 8.44326627                  R-square      =   0.6804
-----------+------------------------------                 Adj R-square  =   0.6760
     Total | 3882.91938    149 26.0598616                  Root MSE      =   2.9057


------------------------------------------------------------------------------------
         Y |     Coef.   Std. Err.        t     P>|t|      [95% Conf. Interval]
-----------+------------------------------------------------------------------------
        X1 |  1.853117   .2721937     6.808     0.000      1.315198    2.391035
        X2 |  2.990529   .2745429    10.893     0.000      2.447968    3.53309
     _cons |  9.033703   .2379768    37.960     0.000      8.563406    9.504001
------------------------------------------------------------------------------------
```

We defined the true model:

$$E[Y] = 9 + 2X1 + 3X2 \qquad\qquad R^2 = .70$$

From this particular $n = 150$ artificial sample, **regress** obtained reasonably good estimates:

$$Yhat = 9.03 + 1.85X1 + 2.99X2 \qquad\qquad R^2 = .68$$

Some samples generated in this manner will less closely resemble the underlying model. Monte Carlo simulations create many samples and record the long-run behavior of statistical procedures that estimate model parameters. By building regression models into a simulation like Example 2, we could compare the performance of **regress** with **qreg**, for instance.

## Bootstrapping

Bootstrapping refers to a process of repeatedly sampling, with replacement, from the data at hand. Instead of trusting theory to describe the sampling distribution of an estimator, we approximate that distribution empirically. Drawing $B$ bootstrap samples of size $n$ (from an original sample also size $n$) obtains $B$ new estimates. Their distribution forms a basis for standard errors or confidence intervals (Efron and Tibshirani 1986; for an introduction see Stine in Fox and Long 1990). Bootstrapping seems most attractive in situations where the estimator is theoretically intractable, or where the usual theory rests on untenable assumptions.

Unlike Monte Carlo simulations, which fabricate their data, bootstrapping often works from real data. For illustration, consider these data (from Zuppan 1973) on 21 New York counties:

```
. use c:\stustata\newyork, clear
. describe

Contains data from c:\stustata\newyork.dta
  Obs:      21 (max=   2620)                   from Jeffrey Zupan, 1973
  Vars:      8 (max=     99)
 Width:     30 (max=    200)
    1. county       str11   %11s        New York county
    2. X            float   %9.0g       people per square mile
    3. Y            float   %9.0g       NOx emissions tonnes/mile^2
    4. hc           float   %9.0g       HC emissions tonnes/mile^2
    5. co           float   %9.0g       CO emissions tonnes/mile^2
    6. low          byte    %8.0g       % w/ <3k 1966 incomes
    7. high         byte    %8.0g       % w/ >10k 1966 incomes
    8. speed        byte    %8.0g       average speed, mph
Sorted by:
```

The distribution of *X*, people per square mile, is positively skewed:

```
. summarize X, detail
```

```
                      people per square mile
--------------------------------------------------------------
          Percentiles      Smallest
  1%        527.6873        527.6873
  5%        683.7607        683.7607
 10%        781.9706        781.9706      Obs                  21
 25%         1228.07         905.028      Sum of Wgt.          21

 50%        2284.946                      Mean           9660.838
                            Largest       Std. Dev.      15921.45
 75%        7382.813        17110.09
 90%        33690.48        33690.48      Variance        2.53e+08
 95%        38260.87        38260.87      Skewness        2.176967
 99%         61703.7         61703.7      Kurtosis        6.863425
```

Stata's **boot** command performs bootstrap sampling, but before using **boot** we must first write our own program to carry out the analysis we want, and create a 1-observation dataset holding the result. For example, here is such a program (*centers.do*) for the mean and median of *X*:

```
program define centers
     summarize X, detail
     clear
     set obs 1
     gen mean = _result(3)
     gen median = _result(10)
end
```

Assume we have *centers.do* and *newyork.dta* in *c:\stustata*. These commands find the mean and median of *X* (population density) for each of *B* = 100 bootstrap samples:

```
. use c:\stustata\newyork, clear
. run c:\stustata\centers.do
. boot centers, iterate(100) clear
```

```
Bootstrap:
          Program:          centers
          Arguments:
          Options:

          Replications:     100
          Data set size:    21
          Sample size:      _N
```

Calculation proceeds slowly; **boot** draws 100 random samples and repeats *centers.do* for each. Eventually, we obtain a new dataset containing 100 bootstrap means and medians:

```
Contains data
  Obs:     100 (max=  2620)                 centers bootstrap
  Vars:      4 (max=    99)
Width:      16 (max=   200)
   1. _rep        long    %10.0g           replication
   2. mean        float   %9.0g
   3. median      float   %9.0g
   4. _obs        long    %10.0g           observations
Sorted by:
Note:  Data has changed since last save

. summarize
```

```
Variable |      Obs        Mean    Std. Dev.        Min         Max
---------+-----------------------------------------------------------------
    _rep |      100        50.5    29.01149           1         100
    mean |      100    9150.971    3348.847    2875.344    18188.02
  median |      100    2938.112    1998.144     1228.07    13377.78
    _obs |      100          21           0          21          21
```

Graphs dramatize the difference between the bootstrap distributions of mean and median:

`. graph mean median, oneway box title(100 bootstrap samples)`

100 bootstrap samples

**Figure 14.4**

The median exhibits less sample-to-sample variation. Furthermore, the mean and median estimate different parameters. With skewed distributions, the sample median is no longer an unbiased estimator of the population mean.

For a second example, bootstrap the regression of $Y$ (nitrous oxide emissions) on $X$ (population density). In the original sample, crowded counties have worse air pollution:

```
. use c:\stustata\newyork, clear
. regress Y X

    Source |       SS       df       MS                  Number of obs =       21
---------+------------------------------              F(  1,    19) =    63.26
   Model | .163164127        1  .163164127              Prob > F      =   0.0000
Residual | .049008291       19  .002579384              R-square      =   0.7690
---------+------------------------------              Adj R-square  =   0.7569
   Total | .212172418       20  .010608621              Root MSE      =   .05079

-----------------------------------------------------------------------------
       Y |      Coef.   Std. Err.       t     P>|t|      [95% Conf. Interval]
---------+-------------------------------------------------------------------
       X |   5.67e-06   7.13e-07     7.953   0.000      4.18e-06    7.17e-06
   _cons |   .0700557   .0130504     5.368   0.000       .042741    .0973705
-----------------------------------------------------------------------------
```

The usual $t$ and $F$ tests rest on assumptions that seem implausible here, however. As an alternative, we might try bootstrapping to estimate standard errors.

To do this, we need to write a small program like *lines.do*:

```
program define lines
     regress Y X
     clear
     set obs 1
     gen constant = _b[_cons]
     gen slope = _b[X]
end
```

Then type:

```
. use c:\stustata\newyork, clear
. run c:\stustata\lines.do
. boot lines, iterate(100) clear

Bootstrap:
          Program:              lines
          Arguments:
          Options:

          Replications:        100
          Data set size:       21
          Sample size:         _N

Contains data
  Obs:     100  (max=   1473)                  lines bootstrap
  Vars:      4  (max=     99)
Width:      16  (max=    200)
  1. _rep            long     %10.0g            replication
  2. constant        float    %9.0g
  3. slope           float    %9.0g
  4. _obs            long     %10.0g            observations
Sorted by:
Note:  Data has changed since last save

. summarize

Variable |       Obs        Mean    Std. Dev.         Min         Max
---------+-----------------------------------------------------------
    _rep |       100        50.5     29.01149           1         100
constant |       100    .0678776      .011872    .0367667    .1026296
   slope |       100    6.27e-06    1.88e-06    4.45e-06    .0000156
    _obs |       100          21            0          21          21
```

The mean bootstrap-sample slope ($6.27 \times 10^{-6}$) is somewhat higher than the original-sample slope ($5.67 \times 10^{-6}$). The bootstrap standard deviation ($1.88 \times 10^{-6}$) implies much greater variation than the original-sample standard error estimate ($7.13 \times 10^{-7}$) does. Figure 14.5 shows the distribution of bootstrap slope estimates, a far cry from the symmetrical, bell-shaped sampling distributions commonly pictured in textbooks:

```
. graph slope, bin(12) xlabel ylabel b2(bootstrap slope)
```

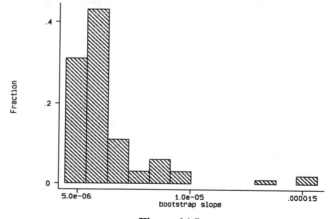

**Figure 14.5**

`boot` performs "data resampling," which resamples entire cases. Data resampling does not require assuming fixed $X$ and independent, identically distributed (*i.i.d.*) errors. Consequently it often yields larger standard errors and skewed, multimodal sampling distributions. If the usual assumptions are false, we are right to abandon them, and bootstrapping may provide better guidance. If the assumptions are true, on the other hand, data resampling is too pessimistic.

Since it scrambles the case sequence, data resampling is also inappropriate with time or spatial series. We could get bootstrap time series in which 1969 appears three times, and 1976 not at all, for instance.

Residual resampling, an alternative regression bootstrap approach, retains the fixed-$X$ and *i.i.d.*-errors assumptions. Residuals from the original-sample regression, divided by $\sqrt{1-K/n}$, are resampled and added to original-sample predicted $Y$ values to generate bootstrap $Y^*$ values, that we then regress on original-sample $X$ values. Residual resampling, preferred for many regression applications, is difficult using `boot`. See Hamilton (1991b) for examples of programs performing residual resampling, or embedding a bootstrap within a Monte Carlo experiment.

The bootstrap's growing popularity derives partly from optimism; its actual performance sometimes disappoints. Monte Carlo simulation provides a way to evaluate bootstrapping objectively. The simulation generates samples according to a user-designed model; we then apply bootstrapping to discover (for example) how often bootstrap-based confidence intervals contain the true model parameters. Experiments along these lines, which require at least four million iterations per model/sample size (2,000 bootstrap resamplings for each of 2,000 Monte Carlo samples), help to delineate the conditions under which bootstrapping succeeds.

## Also Type `help`

| | |
|---|---|
| `ado` | ado-file manipulation |
| `boot` | bootstrap data resampling |
| `capture` | capture return code programming command |
| `display` | display strings and values of scalar expressions |
| `do` | execute commands from a file |
| `egen` | extended `generate` |
| `generate` | create variables from mathematical or logical expressions |
| `if` | programming command |
| `log` | write commands and text output (as ASCII file) to printer or disk |
| `more` | pause until key is depressed |
| `obs` | increase number of observations in dataset |
| `parse` | parse program arguments |
| `program` | define and manipulate programs |
| `quietly` | quietly and noisily perform Stata command |
| `rmsg` | error and return messages |
| `sample` | draw a # percent random sample from the data at hand |
| `set` | set system parameters, including random-number seed |
| `version` | Stata version number |
| `while` | programming command |

# References

Aldrich, John H. and Forrest D. Nelson (1986). *Linear Probability, Logit, and Probit Models.* Beverly Hills: Sage.

Baker, Susan P., R. A. Whitfield, and Brian O'Neill (1987). "Geographic variations in mortality from motor vehicle crashes." *New England Journal of Medicine* 316(22):1384–1387.

Beatty, J. Kelly, Brian O'Leary, and Andrew Chaikin, Eds. (1981). *The New Solar System.* Cambridge, MA: Sky.

Belsley, David A., Edwin Kuh, and Roy E. Welsch (1980). *Regression Diagnostics: Identifying Influential Data and Sources of Collinearity.* New York: Wiley.

Brown, L.R., W.U. Chandler, C. Flavin, C. Pollock, S. Postel, L. Starke, and E.C. Wolf (1986). *State of the World 1986.* New York: Norton.

Chambers, John M., William S. Cleveland, Beat Kleiner, and Paul A. Tukey, Eds. (1983). *Graphical Methods for Data Analysis.* Belmont, CA: Wadsworth.

Chen, A.A., A.R. Daniel, S.T. Daniel, and C.R. Gray (1990). "Wind Power in Jamaica." *Solar Energy* 44(6):355–365.

Cleveland, William S. (1985). *The Elements of Graphing Data.* Monterey, CA: Wadsworth.

Computing Resource Center (1992). *Stata Reference Manual.* Santa Monica, CA: Computing Resource Center.

Council on Environmental Quality (1988). *Environmental Quality 1987–1988.* Washington, DC: Council on Environmental Quality.

Danuso, Francesco (1991). "Nonlinear regression command." *Stata Technical Bulletin* 1(May):17–19.

Efron, Bradley and R. Tibshirani (1986). "Bootstrap methods for standard errors, confidence intervals, and other measures of statistical accuracy." *Statistical Science* 1(1):54-77.

Fox, John (1991). *Regression Diagnostics.* Newbury Pack, CA: Sage.

Fox, John and J. Scott Long, Eds. (1990). *Modern Methods of Data Analysis.* Beverly Hills: Sage.

Frigge, Michael, David C. Hoaglin, and Boris Iglewicz (1989). "Some implementations of the boxplot," *The American Statistician,* 43(1):50–54.

Greene, W.H. (1990). *Econometric Analysis.* New York: MacMillan.

Hall, Peter (1988). "Theoretical comparison of bootstrap confidence intervals." *The Annals of Statistics* 16(3):927-953.

Hamilton, Lawrence C. (1985). "Concern about toxic wastes: Three demographic predictors." *Sociological Perspectives* 28(4):463–486.

——————— (1990). *Modern Data Analysis: A First Course in Applied Statistics.* Pacific Grove, CA: Brooks/Cole.

——————— (1991a). "How robust is robust regression?" *Stata Technical Bulletin* 2(July): 21–26.

——————— (1991b). "Bootstrap programming." *Stata Technical Bulletin* 4(November):18–27.

——————— (1992a). *Regression with Graphics: A Second Course in Applied Statistics.* Pacific Grove, CA: Brooks/Cole.

——————— (1992b). "Quartiles, outliers and normality: Some Monte Carlo results" *Stata Technical Bulletin* 6(March):4–5.

Hanushek, Eric A. and John E. Jackson (1977). *Statistical Methods for Social Scientists.* New York: Academic Press.

Hilbe, Joseph (1991). "Data format conversion using DBMS/Copy and Stat/Transfer." *Stata Technical Bulletin* 3(September):3–6.

——————— (1992). "Stat/Transfer review update." *Stata Technical Bulletin* 6(March):3.

Hoaglin, David C., Boris Iglewicz, and John W. Tukey (1986). "Performance of some resistant rules for outlier labeling." *Journal of the American Statistical Association* 81(396): 991–999.

Hosmer, David W. and Stanley Lemeshow (1989). *Applied Logistic Regression.* New York: Wiley.

Johnston, J. (1984). *Econometric Methods*, 3rd edition. New York: McGraw-Hill.

Judson, D.H. (1992) "Performing loglinear analysis of cross-classifications." *Stata Technical Bulletin* 6(March):7–17.

MacKenzie, Donald (1990). *Inventing Accuracy: A Historical Sociology of Nuclear Missile Guidance.* Cambridge, MA: MIT Press.

Nash, James and Lawrence Schwartz (1987). "Computers and the writing process." *Collegiate Microcomputer* 5(1):45–48.

*Report of the Presidential Commission on the Space Shuttle Challenger Accident* (1986). Washington, DC.

Royston, Patrick (1991). "Lowess smoothing." *Stata Technical Bulletin* 3(September):7–9.

——————— (1992). "Nonlinear regression command." *Stata Technical Bulletin* 7(May):11–18.

Simon, Julian L. and Peter Bruce (1991a). "Resampling: A tool for everyday statistical work." *Chance* 4(1):22–32.

——————— (1991b). "Reply to Boomsma and Molenaar." *Chance* 4(4):30–31.

Tukey, John W. (1977). *Exploratory Data Analysis.* Reading, MA: Addison Wesley.

Werner, Al (1990). "Lichen growth rates for the northwest coast of Spitsbergen, Svalbard." *Arctic and Alpine Research* 22(2):129–140.

World Bank (1987). *World Development Report 1987.* New York: Oxford University Press.

Zupan, Jeffrey M. (1973). *The Distribution of Air Quality in the New York Region.* Baltimore: Johns Hopkins University Press.

# Index

# ORDER FORM
## for IBM users

If you have purchased *Statistics for Stata® 3* for IBM computers, the envelope in this book contains two 3 1/2" IBM diskettes. If your IBM computer requires 5 1/4" diskettes, you may use a computer that has both sizes of disk drive to copy the contents of the enclosed diskettes onto two high-density diskettes (1.2 MB) and install.

If you are unable to copy your disks, complete and mail this form, and we will send you *Student Version of Stata® 3* on 5 1/4" double-density (360K) diskettes.

Please send me a copy of *Student Version of Stata® 3* on 5 1/4" diskettes, ISBN 0-534-18921-0.

**SHIP TO:**

Name    _____

Address    _____

                    _____

                    _____

YOU MUST ENCLOSE YOUR ORIGINAL 3 1/2" DISKS.

# STUDENT  STATA® 3
## Software Registration Card

Registered users may be notified of new product information and may receive information about products and services related to the purchase of STATA® 3.

Check if you are an INSTRUCTOR _____ STUDENT _____ OTHER _____

Name (please print)_____

Permanent mailing address _____

City/State _____ Zip _____

Phone Number  (_____) _____

School Affiliation_____

*Statistics with Stata®3* is REQUIRED_____ RECOMMENDED_____ for the following Course(s)_____

Which disk format did you purchase? IBM 5 1/4" _____ IBM 3 1/2" _____ Macintosh_____

Date of Purchase: _____/_____/_____.

**If you are an instructor, please check correct answers below:**

How did you hear of *Statistics with Stata® 3*? Brochure _____ Convention _____

        Sales Rep _____ Colleague _____ Required for class _____

Where did you obtain this product?

        Campus Bookstore _____ Off-Campus Bookstore _____Desk Copy_____

I have access to personal computers in (check all that apply)

home _____dormitory _____general _____lab _____departmental lab _____work _____

What journals, magazines, and newspapers do you read? _____

_____

Would you like to receive a catalog with other Duxbury statistical products?  yes _____ no _____

**If you are a student, please check correct answers below:**

What year? (circle one)  1  2  3  4  Grad

What is your major/course of study? _____

Do you own a computer?_____Make/Model/Memory _____

_____

What is your age group?  18–21 _____  22–25 _____  26–30 _____  31–35 _____36–40 _____

        41–45 _____  46–50 _____ over 51 _____

What journals, magazines,and newspapers do you read?_____

# Student Stata® Software License Agreement

NOTICE TO USERS:  Please read this notice carefully.  Do not open the package containing the diskettes until you have read the licensing agreement.

## LICENSE

This software is protected by United States copyright law and international treaty provisions.  This software may be used by more than one person and it may be used on more than one computer, **but no more than one copy of this software may be used at the same time.**  Just as a book may not be read by more than one person at a time, this software may not be used by more than one person at a time without violating the license.

Duxbury Press authorizes the purchaser to make archival copies of the software for the sole propose of backing-up the software in order to protect the purchaser's investment from loss.

## LIMITED WARRANTY

The warranty for the enclosed disks is for ninety (90) days.  If, during that time, you find defects in workmanship or material, the Publisher will replace the defective item.  Neither Duxbury Press nor Computing Resource Center provide any other warranties, expressed or implied, and shall not be liable for any damages, special, indirect, incidental, consequential, or otherwise.

For warranty service, contact:
Statistical Computing Editor
Duxbury Press
10 Davis Drive
Belmont, CA 94002